Cambridge Primary

ENDORSED BY

CAMBRIDGE
International Examinations

D1796923

Ready to Go Lessons for Maths

Step-by-step lesson plans for Cambridge Primary

Stage 4

Helen Lewis

Series editor: **Paul Broadbent**

HODDER
EDUCATION
AN HACHETTE UK COMPANY

Every effort has been made to trace all copyright holders, but if any have been inadvertently overlooked the Publishers will be pleased to make the necessary arrangements at the first opportunity.

Although every effort has been made to ensure that website addresses are correct at time of going to press, Hodder Education cannot be held responsible for the content of any website mentioned in this book. It is sometimes possible to find a relocated web page by typing in the address of the home page for a website in the URL window of your browser. Websites included in this text have not been reviewed as part of the Cambridge endorsement process.

Hachette UK's policy is to use papers that are natural, renewable and recyclable products and made from wood grown in sustainable forests. The logging and manufacturing processes are expected to conform to the environmental regulations of the country of origin.

Orders: please contact Bookpoint Ltd, 130 Milton Park, Abingdon, Oxon OX14 4SB. Telephone: (44) 01235 827720. Fax: (44) 01235 400454. Lines are open 9.00–5.00, Monday to Saturday, with a 24-hour message answering service. Visit our website at www.hoddereducation.com.

© Helen Lewis 2013
First published in 2013 by
Hodder Education,
An Hachette UK Company
338 Euston Road
London NW1 3BH

Impression number   5 4 3 2 1
Year   2017 2016 2015 2014 2013

All rights reserved. Apart from any use permitted under UK copyright law, the material in this publication is copyright and cannot be photocopied or otherwise produced in its entirety or copied onto acetate without permission. Electronic copying is not permitted. Permission is given to teachers to make limited copies of individual pages marked © Hodder & Stoughton Ltd 2013 for classroom distribution only, to students within their own school or educational institution. The material may not be copied in full, in unlimited quantities, kept on behalf of others, distributed outside the purchasing institution, copied onwards, sold to third parties, or stored for future use in a retrieval system. This permission is subject to the payment of the purchase price of the book. If you wish to use the material in any way other than as specified you must apply in writing to the Publisher at the above address.

Cover illustration by Peter Lubach
Illustrations by Planman Technologies
Typeset in ITC Stone Serif Medium 10/12.5 by Planman Technologies
Printed in Great Britain by CPI Group (UK) Ltd, Croydon, CR0 4YY

A catalogue record for this title is available from the British Library.

ISBN: 978 1444 177619

Contents

Introduction

About the series

Ready to Go Lessons is a series of photocopiable resource books providing creative teaching strategies for primary teachers. These books support the revised Cambridge Primary curriculum frameworks for English, Mathematics and Science at Stages 1–6 (ages 5–11). They have been written by experienced primary teachers to reflect the different teaching approaches recommended in the Cambridge Primary Teacher Guides. The books contain lesson plans and photocopiable support materials, with a wide range of activities and appropriate ideas for assessment and differentiation. As the books are intended for international schools we have taken care to ensure that they are culturally sensitive.

Cambridge Primary

The Cambridge Primary curriculum frameworks show schools how to develop the learners' knowledge, skills and understanding in English, Mathematics and Science. They provide a secure foundation in preparation for the Cambridge Secondary 1 (lower secondary) curriculum. The ideas in this book can also be easily incorporated into existing curriculum frameworks already in your school.

How to use this book

This book covers each of the units of the scheme of work for Mathematics at Stage 4. It can be worked through systematically (as all the learning objectives are covered), or used to support areas where you feel you need more ideas. It is not prescriptive – it gives ideas and suggestions for you to incorporate into your own teaching as you see fit.

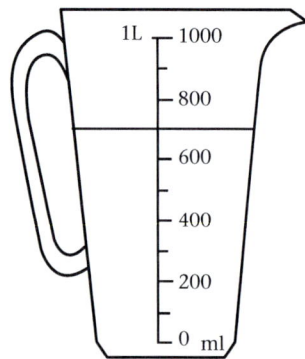

Each step-by-step lesson plan shows you the learning objectives you will cover, the resources you will need and how to deliver the lesson. Each lesson includes a Starter activity, Main activities and a Plenary that draws the lesson to a close and recaps the learning objectives. Success criteria are provided in the form of questions to help you assess the learners' level of understanding. The 'Differentiation' section provides support for the less-able learners and extension ideas for the more able.

For each lesson plan there is at least one supporting photocopiable activity page. At the end of each unit there are also suggestions for assessment activities. Answers to activities can be found at www.hoddereducation.com/~~checkpoint~~extras. *[handwritten: cambridge]*

Learning objectives

The *Mathematics Curriculum Framework* provides a set of learning objectives for each stage. At the start of each lesson you need to re-phrase the learning objectives into child-friendly language so that you can share them with the learners at the outset. It sometimes helps to express them as *We are learning to / about …* statements. This really does help the learners to focus on the lesson's outcomes. For example: 'Understand what each digit represents in three-digit numbers and partition into hundreds, tens and units.' (Stage 3) could be introduced to the learners at the start of the lesson as: *We are learning about the value of each of the digits in a three-digit number.* To avoid unnecessary repetition we have not included such statements at the start of each lesson plan but it is understood that the teacher would do this.

The overview chart on pages 6–7 shows you how the learning objectives are covered in the lessons in this book.

Success criteria

These are the measures that the teacher and, eventually, the learner will be able to use to assess the outcome of the learning that has taken place in each lesson. They are included as a series of questions, which will help you as teacher to assess the learners' understanding of the skills and knowledge covered in the lesson.

Problem-solving skills

Maths teaching is concerned with more than just the learning of mathematical facts. Problem-solving skills are also **essential** and are planned as an on-going and sequential part of each unit.

The activities in these books will show you how to incorporate the practical nature of problem-solving so that it is part of the teaching process. Problem-solving objectives are worked into every teaching unit, with these skills underpinning all other strands to help the learners understand mathematical relationships and functions. These skills need to be used regularly in familiar and new contexts in order for the learners to become mathematical thinkers who are capable of questioning, reasoning and seeking answers through investigation.

The key to successful mathematical problem-solving teaching lies in providing the learners with opportunities to learn by doing, that is, through **active learning**.

Starters

These are an important part of each lesson, consisting of whole-class, teacher-led interactive activities. The purpose varies for each lesson, and can include:

- practice and consolidation of existing skills – often mental calculation but also properties of shape and the language of number

- quick recall – to secure knowledge of number facts and build up speed and accuracy

- revisiting previous learning – to return to aspects of Maths that may have caused difficulty or to strengthen the learners' knowledge and use of mental or written strategies

- preparation for the main part of the lesson – linked to the objectives for the lesson to support the learning.

Formative assessment

Formative assessment is on-going assessment that occurs in every lesson and informs the teacher and learners of the progress they are making, linked to the success criteria. The types of questions to ask that will support teachers in making formative assessments have been incorporated into each lesson in the 'Success criteria' sections.

One of the advantages of formative assessment is that any problems that arise during the lesson can be responded to immediately. Formative assessment influences the next steps in learning and may influence changes in planning and / or delivery for subsequent lessons.

Summative assessment

Summative assessment is essential at the end of each unit of work to assess exactly what the learners know, understand and can do. The assessment sections at the end of each unit are designed to provide you with a variety of opportunities to check the learners' understanding of the unit. These activities can include specific questions for teachers to ask, activities for the learners to carry out (independently, in pairs or in groups) or written assessment.

The information gained from both the formative and summative assessment ideas can then be used to inform future planning in order to close any gaps in the learners' understanding as recommended by *Assessment for Learning* (AFL).

Appropriate use of ICT

At the planning stage teachers need to consider how the use of ICT in a lesson will enhance the learning process. Ensure that the ICT resources you use support and promote the learners' understanding of the learning objectives. Activities included in this book have been designed to be carried out without the need for state-of-the-art ICT facilities. Suggestions have also been included for schools with internet access and / or the use of interactive whiteboards. This is in order to cater for most teachers' needs.

In these lessons the author sometimes asks for the teacher to display an enlarged version of the photocopiable page at the front of the class. We have not specified whether this should be using an overhead projector, interactive whiteboard or flipchart, as schools will have different resources available to them.

We hope that using these resources will give you confidence and creative ideas in delivering the Cambridge Primary curriculum framework.

Paul Broadbent, Series Editor

Overview chart

		Lesson	Framework codes	Page
Term 1	**Unit 1A: Number and problem solving**	Numbers to 1000	4Nn9 4Nn10 4Nn11 4Nn12	8
		Numbers to 10000	4Nn1 4Nn3 4Nn11 4Nn12	11
		Addition strategies 1	4Nc6 4Ps4	13
		Addition strategies 2	4Nc9 4Ps3 4Pt3	15
		Addition strategies 3	4Nc17 4Pt1 4Pt8	17
		Subtraction strategies 1	4Nc10 4Pt4 4Ps9	19
		Subtraction strategies 2	4Nc18 4Nc19 4Ps3	21
		Multiplying and dividing	4Nc5 4Nc15 4Nn8 4Ps9	23
		Multiplication strategies 1	4Nc20 4Ps4 4Ps5	25
		Multiplication strategies 2	4Nc13 4Ps2	27
		Multiplication strategies 3	4Nc21 4Ps1	29
		Multiplication strategies 4	4Nc14 4Nc22 4Ps2	31
		Division strategies 1	4Nc23 4Nc25	33
	Unit assessment			36
	Unit 1B: Measure and problem solving	Length 1	4Ml1 4Ml2 4Ml4	38
		Area and perimeter 1	4Ma1 4Ma2 4Ma3	40
		Mass 1	4Ml1 4Ml2	42
		Capacity 1	4Ml1 4Ml4 4Pt8 4Ps9	44
		Time 1	4Mt1 4Mt2	46
		Time 2	4Pt2 4Ps1	49
		Time 3	4Mt3 4Pt2 4Ps1	51
	Unit assessment			53
	Unit 1C: Handling data and problem solving	Handling data 1	4Dh1 4Ps5 4Ps9	55
		Handling data 2	4Dh1 4Dh2	58
		Handling data 3	4Dh3	60
		Handling data 4	4Dh1 4Ps9	62
	Unit assessment			64

		Lesson	Framework codes	Page
Term 2	**Unit 2A: Number and problem solving**	Place value 1	4Nn3 4Nn4 4Nn7 4Nn9	66
		Place value 2	4Nn2 4Nn6	68
		Odds and evens 1	4Nn15 4Nn16 4Ps8	70
		Negative numbers 1	4Nn13 4Ps4 4Ps5	72
		Number sequences 1	4Nc5 4Ps6 4Nn14	75
		Addition facts	4Nc1 4Nc2	77
		Doubling and halving 1	4Nc16 4Ps9	79
		Addition and subtraction strategies 1	4Nc9 4Nc10 4Pt1	81
		Addition and subtraction strategies 2	4Nc7 4Nc8 4Nc11 4Nc12	83
		More addition	4Ps1 4Nc17 4Pt3 4Pt8	85
		More subtraction	4Nc18 4Nc19 4Ps3	87
		Multiplication and division facts	4Nc4 4Nc13 4Pt5	89
		Multiplication strategies 5	4Nc14 4Nc22 4Ps2	92
		Division strategies 2	4Nc23 4Nc24 4Pt6	94
	Unit assessment			96

Numbers to 1000

Learning objectives

- Round three- and four-digit numbers to the nearest 10 or 100. (4Nn9)
- Position accurately numbers up to 1000 on an empty number line or line marked off in multiples of 10 or 100. (4Nn10)
- Estimate where three- and four-digit numbers lie on empty 0–1000 or 0–10 000 lines. (4Nn11)
- Compare pairs of three- or four-digit numbers, using the < and > signs and find a number in between each pair. (4Nn12)

Resources

Large cards with three-digit numbers; number lines made from photocopiable page 9; large cards made from photocopiable page 10.

Starter

- Shuffle together a set of large cards, each with a three-digit number written on it.
- Hold the cards up one at a time, asking the learners to round the number on the card to the nearest 10 or 100.

Main activities

- Write ten three-digit numbers on the board. Display a blank 0 to 1000 number line made from photocopiable page 9. Ask volunteers to use estimation to position a few of the numbers on the number line.
- Give pairs of learners a blank 0 to 1000 number line made from photocopiable page 9. Ask pairs to place all the numbers from the board onto the number line.
- Give pairs a marked 0 to 1000 number line made from photocopiable page 9, asking them to place the same numbers accurately.
- Ask the learners to compare the positions of the numbers on the two lines. Ask: *Did you get all the numbers in the right order / position on the blank number line? How close were your estimates?*

- Ask the learners to compare pairs of numbers from the number line, write number sentences about them using < and >, and give a number between them, for example 654 > 564; 564 < 654; a number between 564 and 654 is 600.

Plenary

- Hold up large cards made from photocopiable page 10 one at a time. Ask: *Is this number sentence true or false?*
- Ask the learners to put their thumbs up for a true number sentence and put their thumbs down for a false one.

Success criteria

Ask the learners:

- Estimate where 345 should go on this blank number line. How did you work out where to put it?
- Where does 345 go on this marked number line?
- Write down two three-digit numbers. Can you write number sentences with these numbers using the signs < and >?
- What is 173 rounded to the nearest 10?

Ideas for differentiation

Support: Group these learners together and help them with estimation strategies in the second Main activity.

Extension: Ask these learners to work alone rather than with a partner. In the final Main activity, challenge them to write number sentences with three numbers, for example 245 < 452 < 524.

0 to 1000 number lines

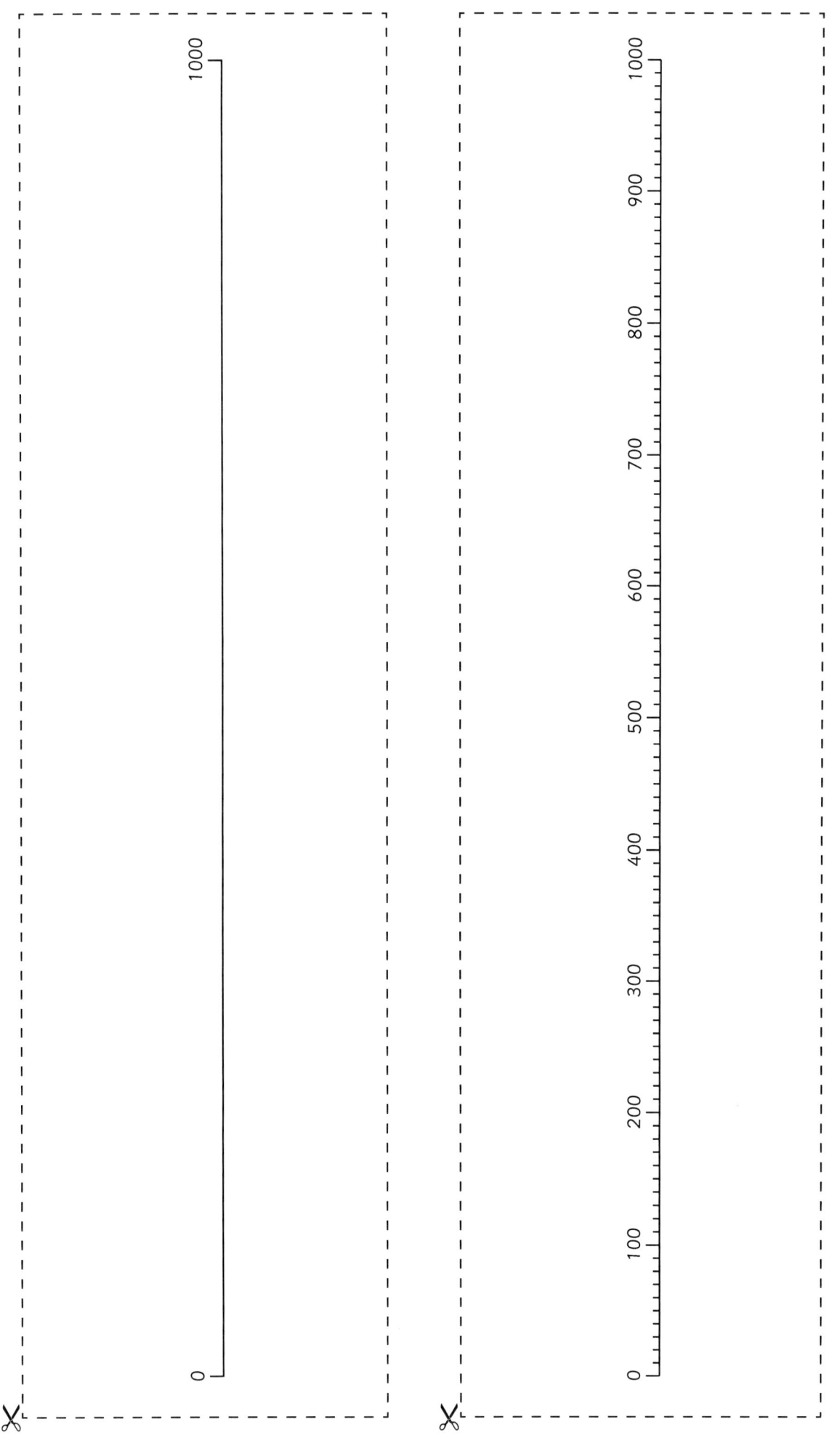

0   100   200   300   400   500   600   700   800   900   1000

0   1000

True or false cards

111 > 158	970 < 478	549 > 502	141 < 481
862 > 548	608 < 615	607 > 646	407 < 223
881 > 271	932 < 629	445 > 205	794 < 555
154 > 849	194 < 306	565 > 472	755 < 224
810 > 777	173 < 437	599 > 949	968 < 904
576 > 455	593 < 403	293 > 341	298 < 834
416 > 244	864 < 482	112 > 866	312 < 345
701 > 204	581 < 114	477 > 621	358 < 871

 Cambridge Primary: Ready to Go Lessons for Maths Stage 4 © Hodder & Stoughton Ltd 2013

Numbers to 10 000

Learning objectives

- Read and write numbers up to 10 000. (4Nn1)
- Understand what each digit represents in a three- or four-digit number and partition into thousands, hundreds, tens and units. (4Nn3)
- Estimate where three- and four-digit numbers lie on empty 0–1000 or 0–10 000 lines. (4Nn11)
- Compare pairs of three- or four-digit numbers, using the < and > signs and find a number in between each pair. (4Nn12)

Resources

Cards made from photocopiable page 12; undivided 0 to 10 000 number lines; sticky notes cut into strips, so that each strip is sticky at one end.

Starter

- Write a four-digit number on the board, for example 4358. Ask the learners to read the number aloud, and then write an addition that will partition the number, for example 4000 + 300 + 50 + 8. Include some numbers with a 0 in them, for example 8075 = 8000 + 70 + 5 or 2907 = 2000 + 900 + 7.

Main activities

- Draw a blank 0 to 10 000 number line on the board. Ask volunteers to estimate where various four-digit numbers lie on the line. Compare pairs of numbers, asking the learners to write a number between them and write a number sentence using the < or > signs.
- Organise the learners into groups of four. Give each group a set of cards made from photocopiable page 12, a number line and some sticky-note strips.
  - Shuffle the cards, deal four to each player and place the remaining four cards face up to form a target number. The players should indicate this number on the number line with a sticky strip.

- The players must arrange their cards to form a number as close as possible to the target number, then indicate their number on the number line with a sticky strip.
- The player who gets closest to the target number scores a point.
- Shuffle the cards and play again.
- The winner is the player with the most points when the time is up.

Plenary

- Ask the learners to describe the strategies they used when playing the game.
- Challenge them to read and write numbers above 10 000. Do this by repeating the Starter activity, using five-digit numbers instead of four-digit numbers.

Success criteria

Ask the learners:

- Can you write six thousand and fifty-three in figures?
- Can you write an addition that will partition the number 3084?
- Where does 8248 go on this blank number line? How did you work out where to put it?
- Use number cards to make two four-digit numbers. Can you say and write a number between them?

Ideas for differentiation

Support: In the game, ask groups of these learners to find the target number by turning over a single card, for example 4 gives a target number of 4000.

Extension: In the game, give groups of these learners extra cards so that each player has six cards.

Number cards

0	1	2	3
4	5	6	7
8	9	0	1
2	3	4	5
6	7	8	9

 Cambridge Primary: Ready to Go Lessons for Maths Stage 4 © Hodder & Stoughton Ltd 2013

Addition strategies 1

Learning objectives

- Add three or four small numbers, finding pairs that equal 10 or 20. (4Nc6)
- Explore and solve number problems and puzzles. (4Ps4)

Resources

One set of large 0 to 9 digit cards; one set of large 10 to 19 number cards; photocopiable page 14; 0 to 20 number lines.

Starter

- Hold up the large 0 to 9 digit cards one at a time, asking the learners to call out the number you need to add in order to make 10, for example if you show 6, the learners should call out 'Four!' Keep the pace brisk.

- Hold up the large 10 to 19 number cards one at a time, asking the learners to call out the number you need to add in order to make 20, for example if you show 12, the learners should call out 'Eight!' Keep the pace brisk.

Main activities

- On the board, write three small numbers (two of which total 10 or 20). Ask the learners to find the total. Discuss strategies used, emphasising finding pairs of numbers that total 10 or 20, adding them first, and then adding the final number. Do several calculations involving three numbers, and then extend to four numbers.

- Display photocopiable page 14. Ask: *How many different ways could Jake make a total of 20 by throwing three darts?* Ask some volunteers to suggest ways of making 20 using three darts. Discuss possible recording methods. Ask: *How will you know when you've found all the possible ways?* Discuss the importance of taking a systematic approach.

- Give each learner photocopiable page 14, asking them to carry out the investigation and record all the different ways they can find of making 20 with three darts.

Plenary

- Ask the learners to share the results of the investigation. Ask: *How many ways are there? Have you found all the possible ways? How do you know?*

- Ask the learners to describe any patterns they can see in the numbers.

Success criteria

Ask the learners:

- What is the total of 7, 9 and 3? How did you work out the answer?
- What is the total of 9, 14 and 6? How did you work out the answer?
- What did you do in the investigation?
- What did you find out in the investigation?

Ideas for differentiation

Support: During the investigation, give these learners materials to support their calculations, for example 0 to 20 number lines.

Extension: Challenge these learners once they have completed the main investigation to look at ways of making a total of 20 using four darts instead of three.

Dartboard investigation

Jake has thrown three darts to make a total of 20.

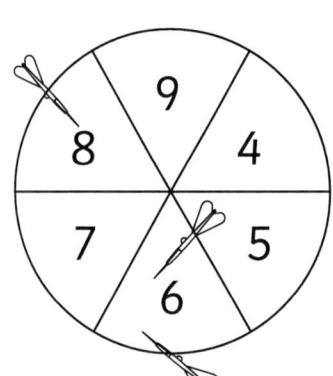

How many different ways could Jake make a total of 20 by throwing three darts? Write your answers on a separate page.

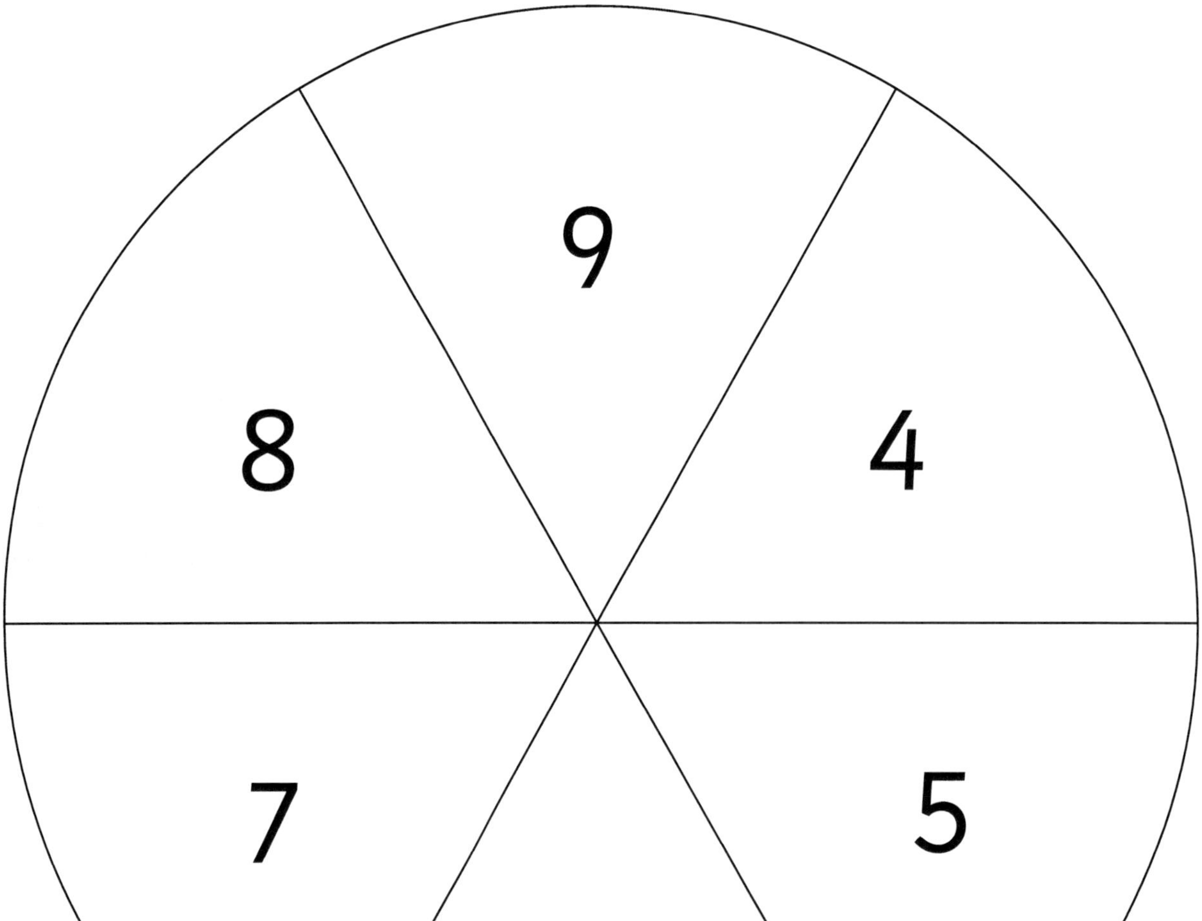

9

8

4

7

5

6

 Cambridge Primary: Ready to Go Lessons for Maths Stage 4 © Hodder & Stoughton Ltd 2013

Addition strategies 2

Learning objectives

● Add any pair of two-digit numbers, choosing an appropriate strategy. (4Nc9)

● Choose strategies to find answers to addition or subtraction problems; explain and show working. (4Ps3)

● Check the results of adding numbers by adding them in a different order or subtracting one number from the total. (4Pt3)

Resources

Number fans made from photocopiable page 16; large 0 to 9 digit cards; large multiples of 10 number cards (10–90); ten-sided dice (made from photocopiable page 18); timers.

Starter

• Before the lesson, make the number fans. Copy photocopiable page 16 onto card. Punch a hole through each circle. Thread the numbers in order onto a loop of wool or string.

• Give each learner a 0 to 9 number fan. Hold up two 0 to 9 digit cards. Ask: *What is the total of these two numbers?* Ask the learners to show the total on their number fan. (Do not show pairs of numbers that total 11, as this number cannot be made on the number fans.) Keep the pace brisk.

• Repeat the activity, holding up two multiples of 10 number cards. (Do not show pairs of numbers that total 110, as this number cannot be made on the number fans.)

Main activities

• Roll a ten-sided dice four times to generate two two-digit numbers, and write the numbers on the board.

• Ask the learners to add the numbers together. Discuss the range of strategies used, for example partitioning and recombining or counting up on a number line. Model each strategy, including jottings, as appropriate.

• Ask: *How could you check your answer?* Discuss the learners' suggestions. Include adding the numbers in a different order and subtracting one number from the answer. Model each strategy, including jottings, as appropriate.

• Organise the learners into groups. Give each group a ten-sided dice and a timer. Ask the learners to roll the dice to generate two two-digit numbers, and set the timer for one minute. All the players must add the numbers within the time limit, and then check the answer. Players with the correct answer get a point. The winning player in each group is the one with the most points at the end of the game.

Plenary

• Write two three-digit numbers on the board, totalling less than 1000. Ask the learners to find the total.

• Discuss the strategies used, emphasising that strategies used to add two-digit numbers can easily be adapted to add three-digit numbers.

Success criteria

Ask the learners:

● What strategy would you use to find the total of 76 and 39?

● What is the total of 76 and 39?

● What strategy could you use to check your answer?

● Use the strategy you described to check your answer. Was your answer correct?

Ideas for differentiation

Support: In the final Main activity, group these learners together and ask them to set the timer for a longer time, for example two minutes.

Extension: In the final Main activity, group these learners together and ask them to set the timer for a shorter time, for example 30 seconds.

0 to 9 number fans

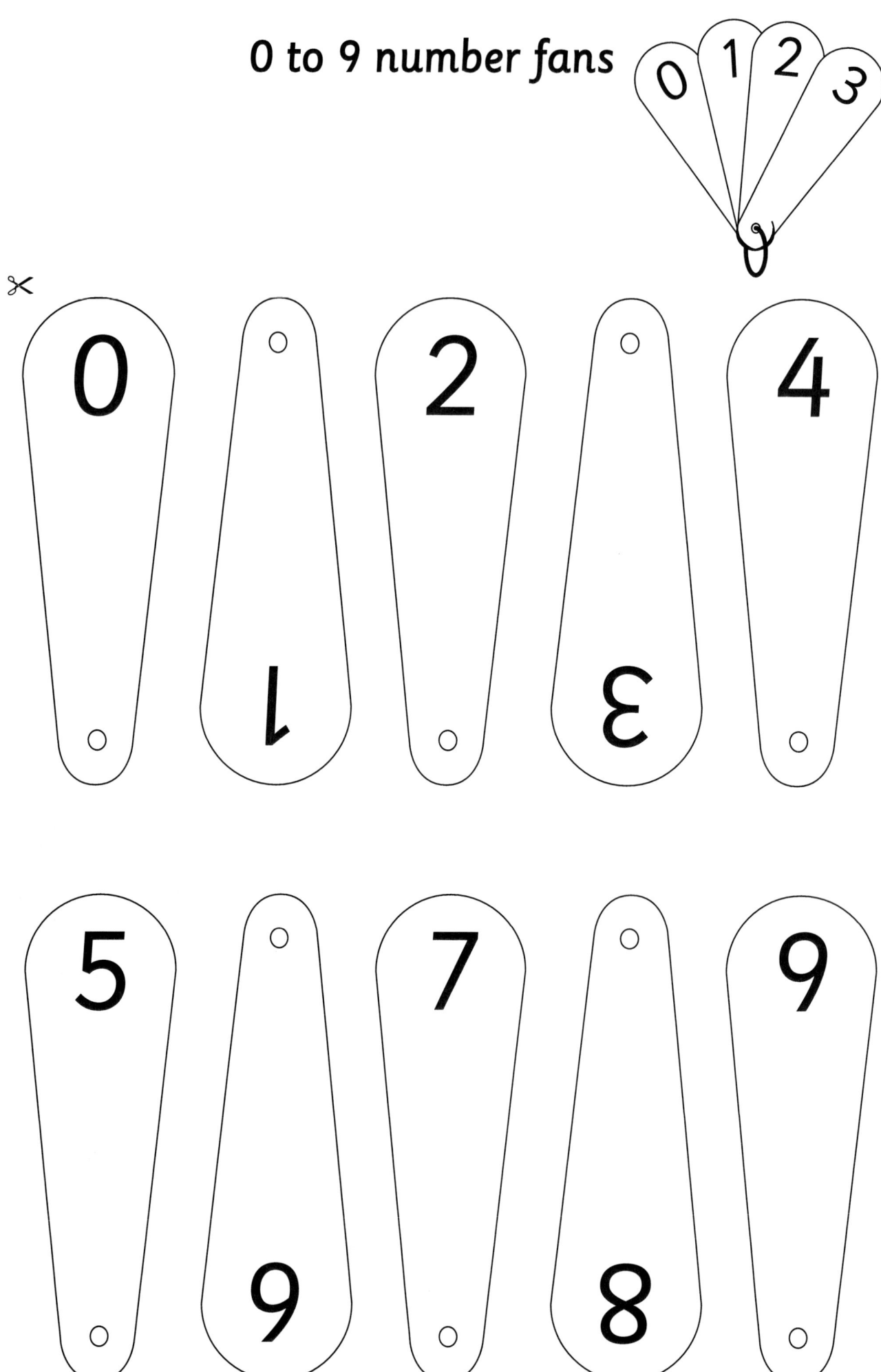

Cambridge Primary: Ready to Go Lessons for Maths Stage 4 © Hodder & Stoughton Ltd 2013

Addition strategies 3

Learning objectives

- Add pairs of three-digit numbers. (4Nc17)
- Choose appropriate mental or written strategies to carry out calculations involving addition and subtraction. (4Pt1)
- Estimate and approximate when calculating and check working. (4Pt8)

Resources

Ten-sided dice (made from photocopiable page 18); timers; calculators.

Starter

- Write a three-digit number on the board and ask the learners to round it to the nearest 100. Repeat, keeping the pace brisk.
- Repeat the activity, this time asking the learners to partition each number into hundreds, tens and units, for example 648 = 600 + 40 + 8.

Main activities

- Write an addition with two three-digit numbers totalling less than 1000 on the board.
- Ask the learners to round both numbers to the nearest 100 and add the two multiples of 100 to give an estimate of the answer. Record the estimate.
- Ask the learners to perform the addition using the non-rounded numbers. Discuss the mental and written strategies used, for example partitioning and recombining or counting up on a number line.
- Check the answer by comparing it to the estimate. Ask: *Is our answer reasonably close to our estimate?*
- Repeat the process for two three-digit numbers totalling more than 1000.

- Organise the learners into groups. Give each group a ten-sided dice, a timer and a calculator. Ask the players to roll the dice to generate two three-digit numbers, and set the timer for two minutes. The players should make an estimate first, then perform the calculation and finally check their answer against the estimate. When the time is up, one player can check the answer on a calculator. Players with the correct answer get a point. The winning player in each group is the one with the most points at the end of the game.

Plenary

- Write a word problem on the board that involves adding a pair of three-digit numbers, for example: *In a certain school there are 475 children in Year 4 and 348 children in Year 5. How many children are there altogether in Years 4 and 5?* Ask the learners to solve the problem and then check their answer.
- Ask the learners to describe the strategies they used to solve the problem and check their answer.

Success criteria

Ask the learners:

- Roll the dice to make two three-digit numbers. Estimate the total of the two numbers. How did you reach your estimate?
- What is the total of the two three-digit numbers?
- How did you work out the answer?
- How could you use your estimate to check your answer?

Ideas for differentiation

Support: Group these learners together in the final Main activity. Do not give them a timer, and ask them to work in pairs within the group.

Extension: In the final Main activity, group these learners together. Ask them to set their timer to one minute.

Template for ten-sided dice

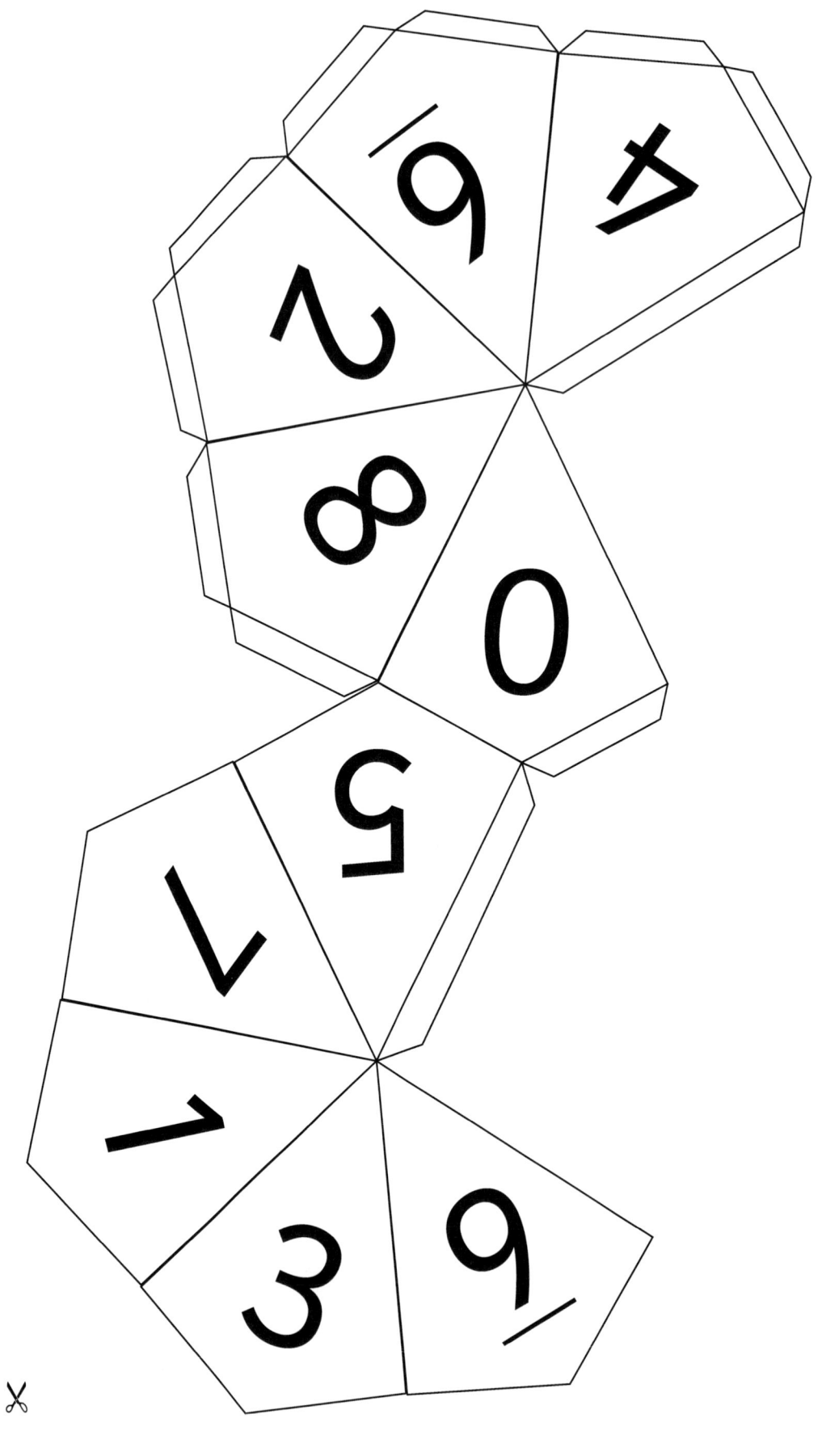

 Cambridge Primary: Ready to Go Lessons for Maths Stage 4 © Hodder & Stoughton Ltd 2013

Subtraction strategies 1

Learning objectives

- Subtract any pair of two-digit numbers, choosing an appropriate strategy. (4Nc10)
- Check subtraction by adding the answer to the smaller number in the original calculation. (4Pt4)
- Explain methods and reasoning orally and in writing. (4Ps9)

Resources

Counting stick; ten-sided dice (see photocopiable page 18); laminated number lines made from photocopiable page 20; dry-wipe pens; cloths; timers.

Starter

- Using a counting stick, lead the learners in:
  - counting on and back in 1s from any two-digit number
  - counting on and back in 10s from any two-digit number.

Main activities

- On the board, sketch a 0 to 100 number line, with multiples of 10 labelled.
- Ask a volunteer to roll a ten-sided dice four times to generate two two-digit numbers. Write the numbers on the board.
- Ask another volunteer to position each number on the number line.
- Ask the learners to find the difference between the numbers. Discuss the range of strategies used, for example counting up or down, counting in 10s and then 1s, or using the multiples of 10 as stepping stones. Ask: *How could we use addition to check the answer?* (Add the answer to one of the original numbers. If the answer is correct, the total should equal the other original number.) Ask the learners to use addition to check the answer.
- Generate more pairs of two-digit numbers, asking the learners to find the difference and then check their answer using addition.

- Organise the learners into groups of four.
  - Give each group a timer and a ten-sided dice.
  - Ask the players to take it in turns to roll a ten-sided dice to generate two two-digit numbers, and then find the difference between them within an agreed time limit, for example two minutes. A player who gives the correct answer within the time limit scores a point. (All the other players can check the active player's answer using addition.)
  - The winner in each group is the player with the most points when each player has had an agreed number of turns.

Plenary

- Write the following pairs of numbers on the board: 93, 68; 69, 42; 81, 55.
- Ask: *Which pair of numbers has the largest difference? How do you know your answer is correct?*

Success criteria

Ask the learners:

- Roll the dice to make two two-digit numbers. What is the difference between them?
- How did you work out your answer?
- What addition could you do to check your answer?
- Do the addition. Is your answer correct?

Ideas for differentiation

Support: In the game, give these learners a laminated number line made from photocopiable page 20, a dry-wipe pen and a cloth.

Extension: In the game, group these learners together. Give players the option of performing the calculation without any jottings to win an extra turn in the game.

0 to 100 number line

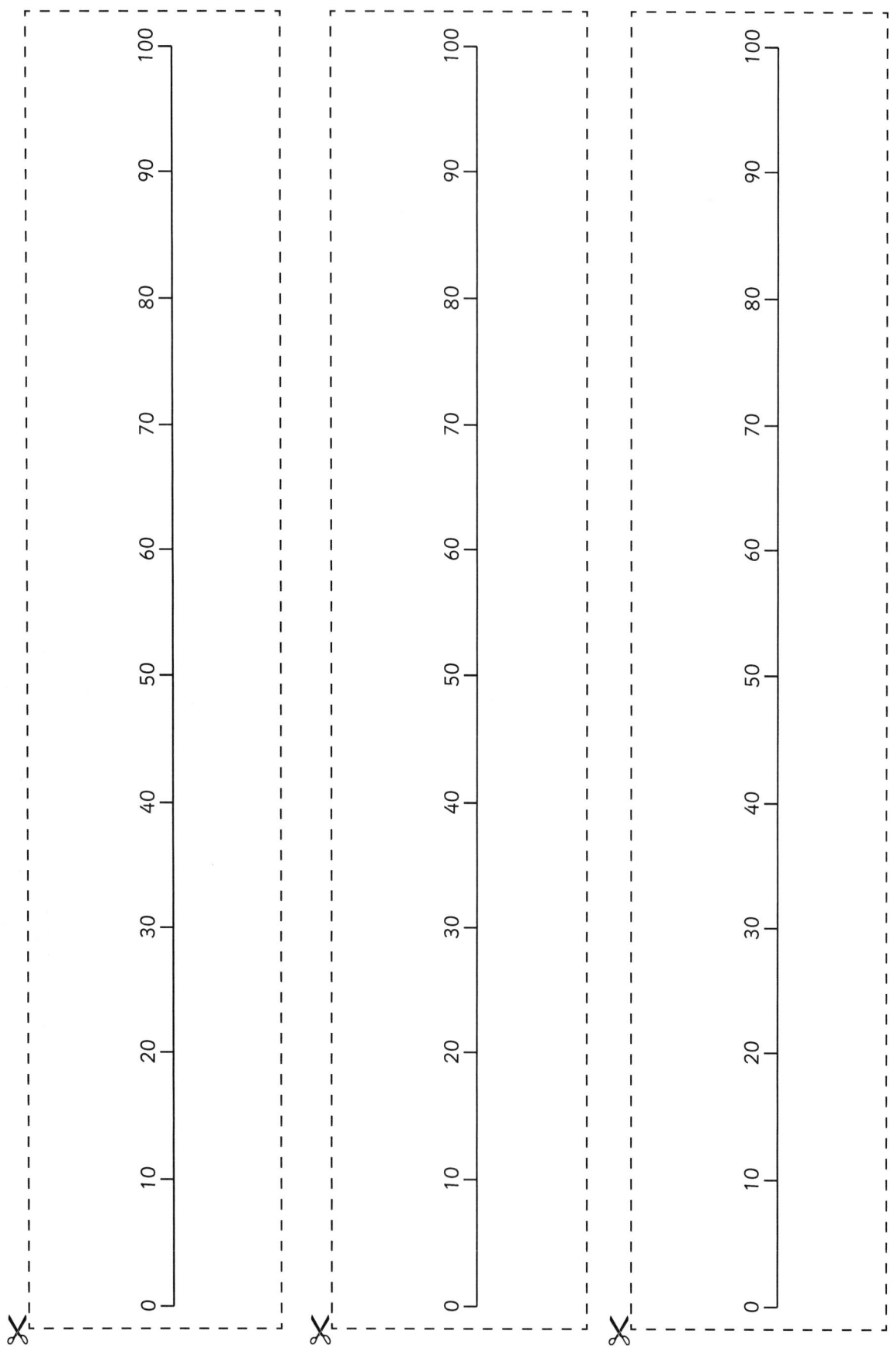

Cambridge Primary: Ready to Go Lessons for Maths Stage 4 © Hodder & Stoughton Ltd 2013

Subtraction strategies 2

Learning objectives

- Subtract a two-digit number from a three-digit number. (4Nc18)
- Subtract pairs of three-digit numbers. (4Nc19)
- Choose strategies to find answers to addition or subtraction problems; explain and show working. (4Ps3)

Resources

Two sets of large 0 to 9 digit cards; two sets of large multiples of 10 number cards (10–90); two sets of large multiples of 100 number cards (100–900); photocopiable page 22; ten-sided dice.

Starter

- Hold up two large 0 to 9 digit cards and ask the learners to say the difference between the two numbers. Repeat until all the 0 to 9 cards have been used.
- Repeat for the large multiples of 10 cards.
- Repeat for the large multiples of 100 cards.

Main activities

- Write the following subtractions on the board one at a time. For each subtraction, ask the learners to work out the answer first, then discuss the range of strategies used, modelling the methods given below.
  - 675 – 432: Model the strategy of partitioning into hundreds, tens and units, subtracting and recombining.
  - 869 – 73: Before the learners attempt this, ask: *Would partitioning and recombining be a good strategy to use for this calculation? Why not?* (Because there aren't enough tens in the larger number to take away the right number of tens in the smaller number.) Model adjusting 869 to 870, subtracting 70, then 3, and finally readjusting by subtracting 1. Ask the learners to check the answer using addition.
  - 404 – 385: Model counting up on a number line.

- Give each learner photocopiable page 22. Organise the learners into pairs. Ask them to answer one question at a time on their own, recording their workings on photocopiable page 22, and continuing on the back of the page if necessary. Ask partners to compare answers and describe to each other the strategy they used. Finally, ask them to check each other's answers using an appropriate strategy, before moving on to the next question.

Plenary

- Write on the board: $\square\square\square - \square\square\square = 333$.
- Ask the learners to solve the equation using just the digits 1 to 6, once each. Ask: *Is there more than one solution?* (Yes, there are six: 456 – 123; 465 – 132; 546 – 213; 564 – 231; 645 – 312; and 654 – 321.)

Success criteria

Ask the learners:

- Roll a ten-sided dice to make two three-digit numbers. What is the difference between them?
- How did you work out your answer?
- What calculation could you do to check your answer?
- Do the calculation to check your answer. Is your answer correct?

Ideas for differentiation

Support: Group these learners together and be on hand to offer support with choosing and using an appropriate calculation strategy.

Extension: Ask these learners to write their own subtractions, each with a particular strategy in mind, and give them to a friend to answer.

Name: _____

Subtraction strategies 2

Decide which method you think will be easiest to do the subtractions below: partitioning, rounding or counting on. Circle the name of the method you have chosen, and show your working.

1. 788 – 315

Method: partitioning / rounding / counting on

2. 206 – 89

Method: partitioning / rounding / counting on

3. 902 – 894

Method: partitioning / rounding / counting on

4. 615 – 71

Method: partitioning / rounding / counting on

 Cambridge Primary: Ready to Go Lessons for Maths Stage 4 © Hodder & Stoughton Ltd 2013

Multiplying and dividing

Learning objectives

● Recognise and begin to know multiples of 2, 3, 4, 5 and 10 up to the tenth multiple. (4Nc5)

● Understand the effect of multiplying and dividing three-digit numbers by 10. (4Nc15)

● Recognise the multiples of 5, 10 and 100 up to 1000. (4Nn8)

● Explain methods and reasoning orally and in writing, make hypotheses and test them out. (4Ps9)

Resources

Photocopiable page 24.

Starter

• Explain the concept of a multiple.

• On the board, write: 'Multiple of 2, 3, 4, 5 or 10?' Above, write a number from one or more of these times tables, for example 12. Ask: *Which of these numbers is twelve a multiple of?* (2, 3 and 4.) Repeat.

• On the board, write: 'Multiple of 5, 10 or 100?' Above, write a number between 100 and 1000 that is a multiple of one or more of these numbers, for example 325. Ask: *Which of these numbers is three hundred and twenty-five a multiple of?* (5.) Repeat.

Main activities

• On the board, draw a place value grid with four columns, as below:

Thousands (Th)	Hundreds (H)	Tens (T)	Units (U)

Remind the learners that each column is ten times bigger than the column to the right of it.

• Write a three-digit number in the grid, for example 487, and demonstrate multiplying it by 10 by moving all the digits one place to the left and inserting a 0 in the units column. Explain that it's moving the digits one place to the left that multiplies their values by 10.

• Ask: *Why do you need to add the zero as well?* (So that when the number is written outside a place value chart, the value of each digit is clear. Otherwise the number would still read as 487.) Repeat for several other three-digit numbers.

• Write a three-digit multiple of 10 in the grid, for example 720. Ask the learners to give instructions for dividing the number by 10. (Move each digit one place to the right.) Ask them to write and say the answer. (72.) Repeat for several other three-digit multiples of 10.

• Hand out photocopiable page 24 to each learner and read through the questions together. Give them time to complete it.

Plenary

• Go through the answers to photocopiable page 24.

• Ask some of the learners who have written their own questions for question 9 to share them with the class.

• Challenge the learners to multiply four-digit numbers by 10, and divide four-digit multiples of 10 by 10.

Success criteria

Ask the learners:

● Which numbers is 20 a multiple of?

● Which of these numbers is 650 a multiple of: 5, 10 and 100?

● What is 303 multiplied by 10? How did you work it out?

● What is 1020 divided by 10? How did you work it out?

Ideas for differentiation

Support: Ask these learners to work in pairs to complete photocopiable page 24. Provide them with pre-drawn place value grids.

Extension: Challenge these learners to try dividing three-digit numbers that are not multiples of 10 by 10, for example ask: *What is 563 divided by 10?*

Name: _____

Multiplying and dividing by 10

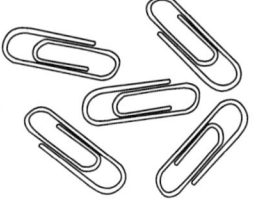

1. How many pens are there in 307 packs of ten pens?

_____

2. How many paper clips are there in ten packets of 144 paper clips?

_____

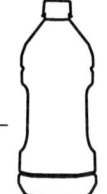

3. How many bottles of apple juice are there in 251 crates of ten bottles?

_____

4. How many stickers are there in ten packets of 120 stickers?

_____

5. There are 3180 dates altogether, divided equally among ten baskets. How many dates are there in each basket?

_____

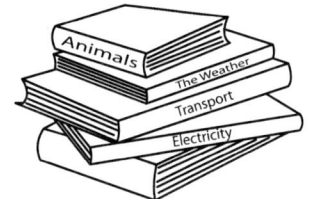

6. There are 2340 pages altogether, divided equally among ten books. How many pages are there in each book?

_____

7. There are 4090 goats altogether, divided equally among ten farmers. How many goats does each farmer have?

_____

8. There are 7560 fish altogether, divided equally among ten nets. How many fish are there in each net?

_____

9. Make up some more questions like these.

_____

_____

_____

  Cambridge Primary: Ready to Go Lessons for Maths Stage 4 © Hodder & Stoughton Ltd 2013

Multiplication strategies 1

Learning objectives

- Double any two-digit number. (4Nc20)
- Explore and solve number problems and puzzles. (4Ps4)
- Use ordered lists and tables to help solve problems systematically. (4Ps5)

Resources

Large 0 to 9 digit cards; large multiples of 10 number cards (10–90); photocopiable page 26.

Starter

- Hold up the large 0 to 9 digit cards one at a time, asking the learners to double each number and write down the answer.
- Repeat for the large multiples of 10 number cards. Ask the learners to relate each doubling fact back to the corresponding 0 to 9 fact, for example 'Double 70 is 140 because double 7 is 14'.

Main activities

- On the board, write a doubling calculation, for example: 'Double 67'. Model using known doubling facts to produce an estimate that gives a range within which the answer must fall, for example double 67 will be between double 60 and double 70 (between 120 and 140).
- Ask the learners to perform the calculation and then check their answer by making sure it is within the range given by the estimate.
- Ask them to describe the strategies they used (for example partitioning into tens and units or doubling the units and the tens separately and then adding the answers).
- Present the learners with more doubling calculations, asking them to make an estimate giving a range within which the answer must fall, perform the calculation and then check their answer against the estimate.
- Hand out photocopiable page 26 and ask the learners to read and solve the puzzle.

Plenary

- Ask: *What is the answer to the puzzle?*
- Ask: *Did anything surprise you about the answer to the puzzle? What surprised you?*
- Ask: *How much would Mira and Amina receive in total if their Grandma doubled the number of days she paid them in Option 2?* (For example if they were paid for 20 days instead of 10.)

Success criteria

Ask the learners:

- Roll two ten-sided dice to make a two-digit number. If you double this number, what two numbers will the answer be between?
- Double the number you made. How did you work out the answer?
- Is your answer reasonable? How do you know?

Ideas for differentiation

Support: For these learners, make the following alterations to photocopiable page 26: change the amount of money offered in Option 1 to 50 c; change the number of days in Option 2 to 7.

Extension: For these learners, make the following alterations to photocopiable page 26: change the amount of money offered in Option 1 to $100; change the number of days in Option 2 to 14.

Name: _____

Doubling puzzle

Mira and Amina are staying with their grandparents. They do lots of jobs around the house, and their Grandma says she will give them some money in return for all their hard work.

Grandma gives Mira and Amina two options for being paid:

- Option 1 – get $5

- Option 2 – get 1c today, double that amount tomorrow, double tomorrow's amount the next day, and so on, doubling the amount every day, for ten days.

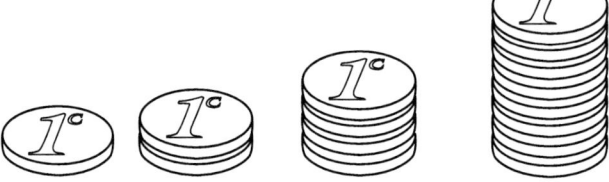

Which option should Mira and Amina choose? Why?
Use the box below to work out your answer.

Multiplication strategies 2

Learning objectives

- Multiply any pair of single-digit numbers together. (4Nc13)
- Explain reasons for a choice of strategy when multiplying or dividing. (4Ps2)

Resources

Large cards made from photocopiable page 28; ten-sided dice made from photocopiable page 18; standard six-sided dice.

Starter

- Use the large cards made from photocopiable page 28 to practise quick-fire recall of the 2, 3 and 5 times tables. Hold up one card at a time and ask the learners to call out the answer to the multiplication.
- Ask a volunteer to come to the front and roll two ten-sided dice to generate two-digit numbers and call the numbers out. Ask the rest of the class to double each number and write down the answer.

Main activities

- Revise methods for multiplying by 4, 6 and 9, for example:
  - ×4 by doubling and doubling again
  - ×6 by multiplying by 3 and doubling the answer
  - ×9 using the following digital sum / finger method: Place the hands side by side, palms up. To calculate three 9s, fold down the third finger (counting from the left). The number of fingers to the left of the folded down finger (2) gives the tens digit of the answer. The number of fingers to the right (7) gives the units digit. So three 9s are 27.
- Demonstrate multiplying by 7, by multiplying by 5 and by 2 and adding the products together.
- Demonstrate this method by drawing an array, for example 6 × 7, and splitting it in two, for example 6 × 5 and 6 × 2. Calculate each product (30 and 12) and add them (42).
- Ask the learners to check the answer by splitting the array in a different way, for example 6 × 4 and 6 × 3.

- Ask the learners how they would multiply by 8, for example by:
  - multiplying by 4, then doubling
  - doubling three times in a row
  - splitting into ×3 and ×5 then adding
  - ×10 and subtract ×2.
  Ask them to calculate 9 × 8 and then check their answer using a different method.
- Organise the learners into pairs and give each pair two ten-sided dice. Ask them to generate two single-digit numbers to multiply, calculating the answer individually. Ask partners to compare answers and describe strategies to each other. Ask them to check each other's answers using a different method.

Plenary

- Write 8 × 12 on the board and ask the learners to work out the answer and then check their answer using a different method. Discuss the range of strategies used.
- Repeat for other multiplications involving single-digit numbers and teens numbers, for example 6 × 13, 9 × 15.

Success criteria

Ask the learners:

- Multiply 9 by 8. How did you work out the answer?
- Why did you choose this method?
- Check your answer by multiplying 9 by 8 using a different method. What answer do you get?
- Which method did you use?

Ideas for differentiation

Support: In the final Main activity, group these learners together and give them standard six-sided dice instead of ten-sided dice.

Extension: In the final Main activity, group these learners together. Ask pairs to choose a calculation they have done, and work together to find as many ways as possible of doing it.

Multiplication cards

0 × 2	2 × 2	4 × 2	6 × 2
7 × 2	8 × 2	9 × 2	10 × 2
1 × 3	3 × 3	4 × 3	5 × 3
6 × 3	7 × 3	8 × 3	9 × 3
0 × 5	3 × 5	4 × 5	5 × 5
6 × 5	7 × 5	8 × 5	10 × 5

Cambridge Primary: Ready to Go Lessons for Maths Stage 4 © Hodder & Stoughton Ltd 2013

Multiplication strategies 3

Learning objectives

- Multiply multiples of 10 to 90 by a single-digit number. (4Nc21)
- Make up a number story for a calculation. (4Ps1)

Resources

Beanbag; large and small multiples of 10 number cards (10–90); large and small 0 to 9 digit cards; photocopiable page 30.

Starter

- Throw a beanbag to a learner while saying a multiplication from the 2, 3, 4, 5, 6, 9 or 10 times tables, for example: *Nine times five.* Choose the question to suit the ability of the learner. The learner must throw the beanbag back, saying the product, for example: 'Forty-five.' Throw the beanbag to another learner. Encourage speed.

Main activities

- Turn over the top card from a pile of large multiples of 10 number cards and the top card from a pile of large 0 to 9 digit cards. Ask the learners to multiply the two numbers together.
- Ask them to give the product and explain the method they used to find it. Relate to known facts; the strategies for multiplying two single-digit numbers together covered in the previous lesson; and a knowledge of place value.
- Display photocopiable page 30. Choose one of the questions from Section A, asking the learners to identify the calculation that needs to be done in order to solve the problem. Solve the problem, and ask the learners to explain how they worked out the answer.

- Write one of the calculations from Section B on the board, asking the learners to make up a number story to go with it. Encourage them to include stories involving measures, for example time, money, distance, mass and capacity. Some possible number stories for 50×7 might be:
  - 'Mahmood cycles 50 km every day for a week. He cycles a total of 350 km.'
  - 'A shelf in a warehouse holds 50 sacks of rice, each weighing 7 kg. The sacks of rice weigh 350 kg altogether.'
  - 'If you earn $7 per hour and you work for 50 hours, you will earn a total of $350.'
- Hand out photocopiable page 30 and ask the learners to complete it.

Plenary

- Ask the learners to give the answers to the questions in Section A on photocopiable page 30, describing their method.
- Ask selected learners to share the number stories they wrote for the questions in Section B.

Success criteria

Ask the learners:

- What is 80×6?
- How did you work out the answer?
- How could you check that your answer is correct?
- Can you make up a number story to go with 80×6?

Ideas for differentiation

Support: In the final Main activity, group these learners together, and work through some extra questions with them. Ask them to complete only the first three questions in each section on photocopiable page 30.

Extension: In the final Main activity, challenge these learners to answer at least one of the questions in Section A entirely mentally, without making any jottings.

Name: _____

Multiplying multiples of 10

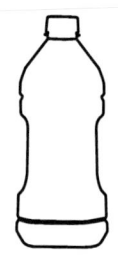

Section A

1. One crate of lemonade contains 80 bottles.
   How many bottles of lemonade are there in two crates?

   _____

2. On the planet Zog, every month has 30 days.
   How many days are there in five months on Zog?

   _____

3. Tia has a paper round. She earns $50 per week.
   How much does she earn in three weeks?

   _____

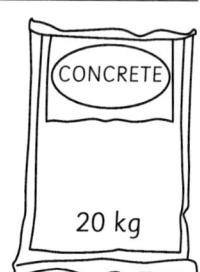

4. The school running track is 60 m long. Samir does four laps
   of the track. How far does he run?

   _____

5. A bag of ready-mix concrete has a mass of 20 kg.
   What is the total mass of seven bags?

   _____

Section B

Write a number story to go with each of these calculations.

1. 30 × 4 _____

   _____

2. 20 × 9 _____

   _____

3. 40 × 8 _____

   _____

4. 90 × 6 _____

   _____

5. 50 × 7 _____

   _____

Cambridge Primary: Ready to Go Lessons for Maths Stage 4 © Hodder & Stoughton Ltd 2013

Multiplication strategies 4

Learning objectives

- Use knowledge of commutativity to find the easier way to multiply. (4Nc14)
- Multiply a two-digit number by a single-digit number. (4Nc22)
- Explain reasons for a choice of strategy when multiplying or dividing. (4Ps2)

Resources

Charts for the 6 and 9 times tables; plain paper; pencils; large and small 0 to 9 digit cards; photocopiable page 32; calculators.

Starter

- Display tables charts for the 6 and 9 times tables. Chant the tables together with the learners.
- Ask them to draw three by three grids, writing a different multiple of 6 or 9 in each square.
- Remove the charts and play a game of times tables bingo. Call out a multiplication from the 6 or 9 times table, for example 5 × 6. Any learners with the product (30) on their grid must circle it. The first learner to correctly circle all the squares on their grid is the winner.

Main activities

- Use three large-digit cards to make one two-digit and one single-digit number. Write the numbers on the board as a multiplication, for example 56 × 8.
- Ask the learners to give an estimate and describe their estimation method.
- On the board, draw a multiplication grid like those on photocopiable page 32. Remind the learners how to use it, by multiplying 56 by 8. First partition 56 into 50 + 6. Explain that because 56 = 50 + 6, you can work out 56 × 8 by working out 50 × 8 and 6 × 8 and adding the answers together. Write 50 and 6 in the top row of the grid and 8 below the multiplication sign, like this:

×	50	6	TOTAL
8			

- Multiply 50 by 8 and write the product in the grid. When multiplying 50 by 8, ask: *How could you work out the answer?* Discuss various strategies.
- Multiply 6 by 8 and write the product in the grid. When multiplying 6 by 8, ask: *Which way round is it easier to multiply these two numbers?* (8 × 6 or 6 × 8?)
- Add the two products together, and write the answer in the 'total' column.
- Give each learner photocopiable page 32. Work through some more calculations together.
- Organise the learners into ability groups of four.
  - Give each group a set of 0 to 9 digit cards.
  - Ask the learners to use the number cards to generate a two-digit by one-digit multiplication, use their multiplication grid to perform the calculation, and then place their grid face down in a single pile.
  - The first correct answer gets 4 points, the second gets 3, the third 2, and the fourth 1.
  - The winner is the player with the most points when the time is up.

Plenary

- Present the learners with a word problem in which they need to multiply a two-digit number by a single-digit number.
- Ask the learners to work in pairs to solve the problem. Then ask them to describe the calculation they did, and the method they used.

Success criteria

Ask the learners:

- Estimate the answer to 39 × 5.
- Do the multiplication. What is the answer?
- Which strategies did you choose to multiply the numbers? Why?

Ideas for differentiation

Support: In the final Main activity, give these learners a calculator for checking answers.

Extension: In the final Main activity, give these learners a set of number cards made from two lots of the numbers 5 to 9.

Name: _____

Multiplication grids

×			TOTAL

×			TOTAL

×			TOTAL

×			TOTAL

Cambridge Primary: Ready to Go Lessons for Maths Stage 4 © Hodder & Stoughton Ltd 2013

Division strategies 1

Learning objectives

- Divide two-digit numbers by single-digit numbers. (4Nc23)
- Understand that multiplication and division are the inverse function of each other. (4Nc25)

Resources

0 to 9 number fans made from photocopiable page 16; large cards made from photocopiable page 34; photocopiable page 35.

Starter

- Give each learner a 0 to 9 number fan.
- Shuffle together a set of large cards made from photocopiable page 34. Hold up one card at a time and ask the learners to show the answer on their number fan. Ask: *How did you work out the answer?* Link to knowledge of multiplication facts, for example $24 \div 6 = 4$ because $4 \times 6 = 24$ (or because $6 \times 4 = 24$).

Main activities

- Display photocopiable page 35. Read the first problem aloud. Ask: *What calculation do you need to do?*
- Ask the learners to suggest how to do the calculation (for example using partitioning). Perform the calculation, and then ask: *How could you check the answer?* For example check that $65 \div 3 = 21$ r2 by doing 21×3 and then adding 2 to see if the answer is 65.
- Re-read the problem, and discuss how to deal with the remainder (whether to round up or down). Ensure that the learners understand why they need to round the answer in the first place (because the question requires a whole number answer), and also why they need to round down in this particular instance (because we're counting full packs, and the leftover chocolate eggs do not make a full pack).

- Work through the second problem in a similar way. This calculation ($57 \div 4$) cannot be solved by simple partitioning into tens and units, and requires a slightly different approach, for example partitioning 57 into $40 + 17$. Ensure that the learners understand why the answer to this question needs to be rounded up (because if Amy were to round down, and buy just 14 packets of cupcakes, she would not have enough cupcakes for the party).
- Hand out photocopiable page 35, asking the learners to work through the rest of the problems in pairs or individually.

Plenary

- Go through the answers to the problems on photocopiable page 35.
- Ask the learners to give each calculation, describe the division strategy they used, say whether they rounded the answer up or down, and explain why.

Success criteria

Ask the learners:

- What is $80 \div 6$?
- How did you work out the answer?
- How could you check your answer?
- If the question the calculation came from was: *A gardener needs 80 tomato seedlings. How many packs of six seedlings does she need?*, what answer would you give? Why?

Ideas for differentiation

Support: Group these learners together and work through a couple more problems on photocopiable page 35 with them.

Extension: Ask these learners to work on photocopiable page 35 individually rather than with a partner. Early finishers could devise similar problems to give to a friend.

Division cards

24 ÷ 4	27 ÷ 3	54 ÷ 9	12 ÷ 6
50 ÷ 5	28 ÷ 4	20 ÷ 2	30 ÷ 6
32 ÷ 8	10 ÷ 5	50 ÷ 2	35 ÷ 7
15 ÷ 3	8 ÷ 2	36 ÷ 4	18 ÷ 9
21 ÷ 7	14 ÷ 2	48 ÷ 6	27 ÷ 9
14 ÷ 7	40 ÷ 8	16 ÷ 4	18 ÷ 3
24 ÷ 3	35 ÷ 5	16 ÷ 8	18 ÷ 2

Cambridge Primary: Ready to Go Lessons for Maths Stage 4 © Hodder & Stoughton Ltd 2013

Division problems

1. Hayley has 65 chocolate eggs to put into packs of three. How many full packs can Hayley make?

2. Fowzia needs 57 cupcakes for a party. There are four cupcakes in a packet. How many packets of cupcakes does Fowzia need to buy?

3. Sofia and Lilly are given $79 to share equally between them. How much money does each girl get?

4. Danah buys a painting for $86. She only has $5 bills. How many $5 bills does Danah need to count out?

5. Aisha wants to put a row of tiles above her bathroom sink. The sink is 81 cm wide. Each tile is 6 cm wide. How many tiles does Aisha need?

6. How many whole weeks are there in 99 days?

7. How many 9 litre buckets can be filled from a water barrel containing 64 litres?

8. Each carriage on Thunder Mountain has eight seats. There are 58 people on the ride. What is the minimum number of carriages that must have people in them?

Unit assessment

Questions to ask

- Can you write a number sentence using two four-digit numbers and the sign >?

- How could you check that this number sentence is correct? 607 − 423 = 184

- Use a set of 0 to 9 digit cards to make two three-digit numbers. What is their total?

- A baker needs 70 eggs to make a large batch of pastries. Eggs come in cartons of six. How many cartons of eggs does the baker need to buy?

- Can you name at least three numbers that 600 is a multiple of?

- Multiply 68 by 7. Make an estimate first, and then multiply. How did you work out the estimate? How did you work out the answer?

Summative assessment activities

Observe the learners while they take part in these activities. You will quickly be able to identify those who appear to be confident and those who may need additional support.

Target 5000

This game assesses the learners' knowledge of numbers and place value up to 10 000.

You will need:

Ten-sided dice; 0 to 10 000 number lines, on which each thousand is labelled and each small unlabelled division represents one hundred; coloured pencils; calculators.

What to do

- Organise the learners into pairs. Give each pair two coloured pencils in different colours, a ten-sided dice, and a 0 to 10 000 number line.

- Ask players to take it in turns to generate four digits by rolling the dice four times. They must use these digits to make the closest possible number to 5000. They should say this number aloud and mark it on the number line using a coloured pencil.

- The game ends when both players have marked ten numbers on the number line. The winner is the player who has made the number closest to 5000. If players cannot decide which number is closest, give them a calculator.

Making 120

This activity assesses the learners' ability to multiply two-digit numbers by single digit numbers, to recognise multiples, and to explore and solve number problems and puzzles.

You will need:

Pencil and paper.

What to do

- Ask: *How many different ways can you make 120 by multiplying two numbers together?*

- If any learners are stuck for a 'way in' to the problem, ask them to start by thinking about different ways of multiplying two numbers together to make 12.

- There are 16 ways of making 120 by multiplying two (whole) numbers: 1×120; 2×60; 3×40; 4×30; 5×24; 6×20; 8×15; 10×12; 12×10; 15×8; 20×6; 24×5; 30×4; 40×3; 60×2; 120×1.

Written assessment

Distribute photocopiable page 37. Ask the learners to read the questions and write the answers. They should work independently.

Name: _____

Number problems

1. Complete these number sentences by writing the sign < or > in each box.

    a)   452 ☐ 398   b)   601 ☐ 579

    c)   999 ☐ 1002   d)   250 ☐ 248

2. Partition each number. Here is an example: 3527 = 3000 + 500 + 20 + 7

    a)   2690 _____   b)   1215 _____

    c)   784 _____   d)   4028 _____

3. For each pair of numbers, find both the sum and the difference.

    a)   13 and 55 _____   b)   89 and 37 _____

         _____   _____

    c)   28 and 67 _____   d)   34 and 82 _____

         _____   _____

4. A crate contains 285 oranges and 167 bananas. How many pieces of fruit does it contain altogether?

    _____

    _____

5. There are 56 tadpoles in a bucket. There are twice as many tadpoles in the pond. How many tadpoles are there in the pond?

    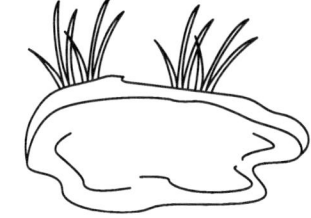

    _____

6. 45 × 4 = _____

7. Henri walks 14 kilometres every day. How many kilometres does he walk in a week?

    _____

8. 48 ÷ 3 = _____

9. Mustafa buys a construction set for $72. He only has $5 bills. How many $5 bills does Mustafa need to count out?

    _____

    _____

Unit 1B: Measure and problem solving

Length 1

Learning objectives

- Choose and use standard metric units and their abbreviations when estimating, measuring and recording length, weight and capacity. (4Ml1)
- Know and use the relationships between familiar units of length, mass and capacity; know the meaning of kilo-, centi-, and milli-. (4Ml2)
- Interpret intervals / divisions on partially numbered scales; record readings accurately. (4Ml4)

Resources

Photocopiable page 39; coloured pens; rulers; metre sticks; tape measures.

Starter

- Draw the following table on the board, and include an additional three rows: one each for centimetre, metre and kilometre. Ask the learners to fill in the missing information.

Unit of length	Abbreviation	How it relates to other units
millimetre		$1\,cm = \square\,mm$ $\square\,mm = 1\,m$

- Ask the learners to work out the meanings of milli-, centi- and kilo-.

Main activities

- Display an enlarged copy of photocopiable page 39 and use a coloured pen to mark off a length on each of the scales. As a class, ask the learners to read each scale and discuss the lengths shown, explaining how they worked them out.
- Draw the following table on the board:

Dimension	Estimate	Measurement

- In the 'Dimension' column, write about a dozen classroom dimensions, ranging from less than 1 cm to more than 1 m, for example the thickness of an exercise book, the height of a chair seat, the length of a window sill.
- Organise the learners into pairs and ask them to copy the table from the board. Hand out photocopiable page 39 and ask the pairs to estimate each dimension on the page using appropriate units, and record their estimates using the abbreviations m, cm and mm.
- Give each learner a ruler. Give each pair of learners a metre stick and a tape measure. Ask the learners to measure the given dimensions using appropriate instruments, and record the measurements using abbreviations.

Plenary

- Ask the learners to share the measurements they have made.
- Ask them which of their estimates were closest to the actual measurements. Discuss whether it is easier to estimate large dimensions or small dimensions, and ask the learners to suggest why this might be the case.

Success criteria

Ask the learners:

- Can you explain the relationships between the metre and the millimetre / metre and centimetre / metre and kilometre?
- What units would you use to measure the height of a door?
- Can you estimate the height of the classroom door?
- Measure the height of the classroom door. Can you record the measurement using standard metric units and their abbreviations?

Ideas for differentiation

Support: In the Main activity, pair these learners with a partner of average ability to support them.

Extension: Ask these learners to calculate the difference between each measurement and its estimate.

Interpreting length scales

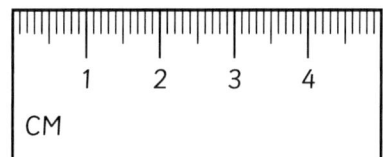

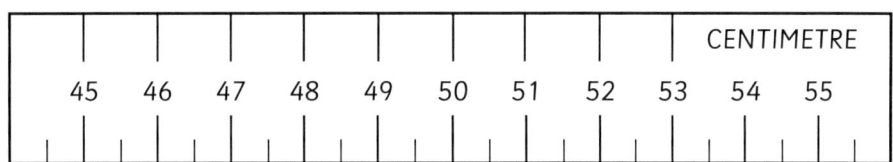

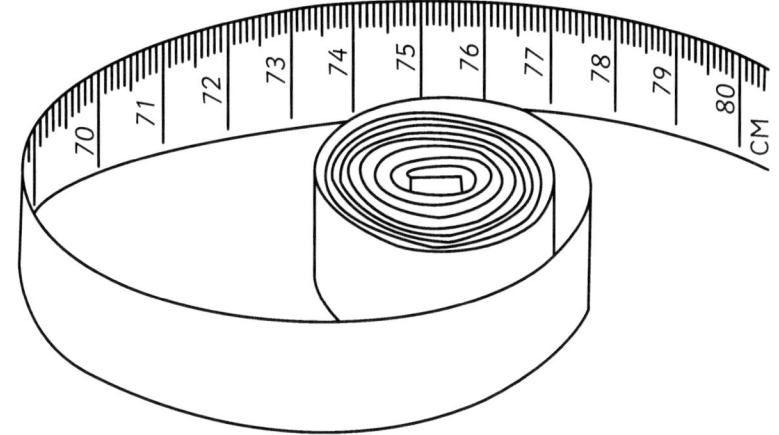

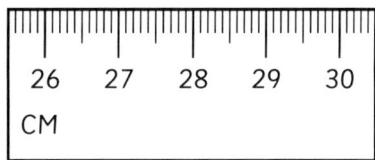

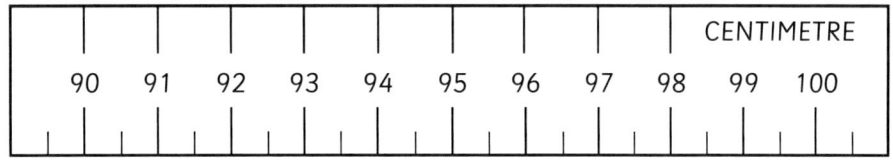

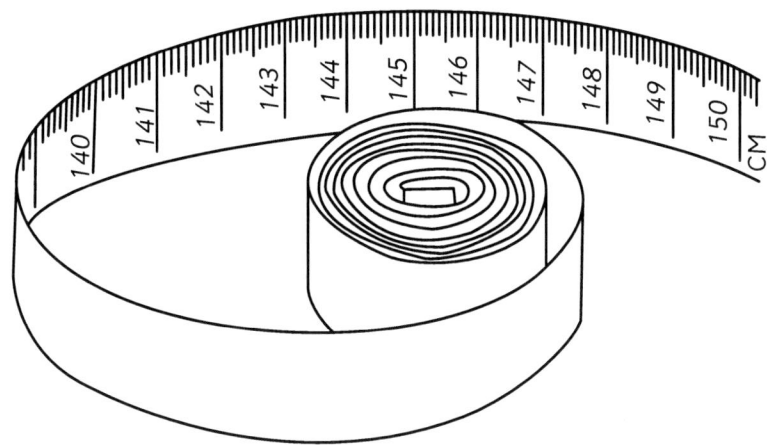

Area and perimeter 1

Learning objectives

- Draw rectangles and measure and calculate their perimeters. (4Ma1)
- Understand that area is measured in square units, e.g. cm squared. (4Ma2)
- Find the area of rectilinear shapes drawn on a square grid by counting squares. (4Ma3)

Resources

Photocopiable page 41; square grid for display; centimetre-squared paper.

Starter

- On the board write an addition calculation that involves adding two doubles, for example double 6 + double 4. Ask the learners to work out the answer as quickly as possible. Repeat for similar calculations, keeping the numbers that are to be doubled below 20.

Main activities

- Display an enlarged copy of photocopiable page 41. Ask: *Which shape is the largest? How could you find out?* (By counting the number of squares it covers.) Explain that the amount of space a 2D shape covers is called its area. Area is measured in square units. Tell the learners that the grid onto which the shapes have been drawn has a grid size of 1 cm. Ask them to work out the area of each shape in square centimetres by counting the number of squares it covers.
- Introduce the term 'perimeter' (the distance around the edge of a shape). Ask the learners to work out the perimeter of each shape in centimetres by counting the number of squares along the edge.
- On the square display grid, draw a rectangle with an area of 24 square centimetres. Ask the learners to find the area and perimeter of the rectangle.

- Challenge the learners to draw (on centimetre-squared paper) as many different rectangles as possible with an area of 24 square centimetres. Ask: *Which rectangle has the longest perimeter? Which rectangle has the shortest perimeter?*

Plenary

- Ask the learners to share what they have found out. There are four rectangles with an area of 24 square centimetres. The rectangle with the longest perimeter is 24 by 1 (perimeter of 50 cm), and the rectangle with the shortest perimeter is 6 by 4 (perimeter of 20 cm).
- Ask the learners to find all the possible rectangles with a perimeter of 16 cm (there are four of them: 1 by 7, 2 by 6, 3 by 5 and 4 by 4.) You may need to remind the learners that a square is a type of rectangle.

Success criteria

Ask the learners:

- On squared paper draw a rectangle with an area of 20 square centimetres. What is its perimeter?
- On squared paper draw a rectangle with a perimeter of 24 cm. What is its area?
- Imagine a rectangle that is 3 cm long and 7 cm wide. What is its area? How did you work it out?
- Imagine a rectangle that is 5 cm long and 4 cm wide. What is its perimeter? How did you work it out?

Ideas for differentiation

Support: Ask these learners to investigate the perimeters of rectangles with an area of 12 square centimetres.

Extension: Ask these learners to investigate the perimeters of rectangles with an area of 48 square centimetres.

Which shape is the largest?

Find the area of each shape in square centimetres by counting the number of squares it covers. Which shape is the largest?

A B C D

E F G H

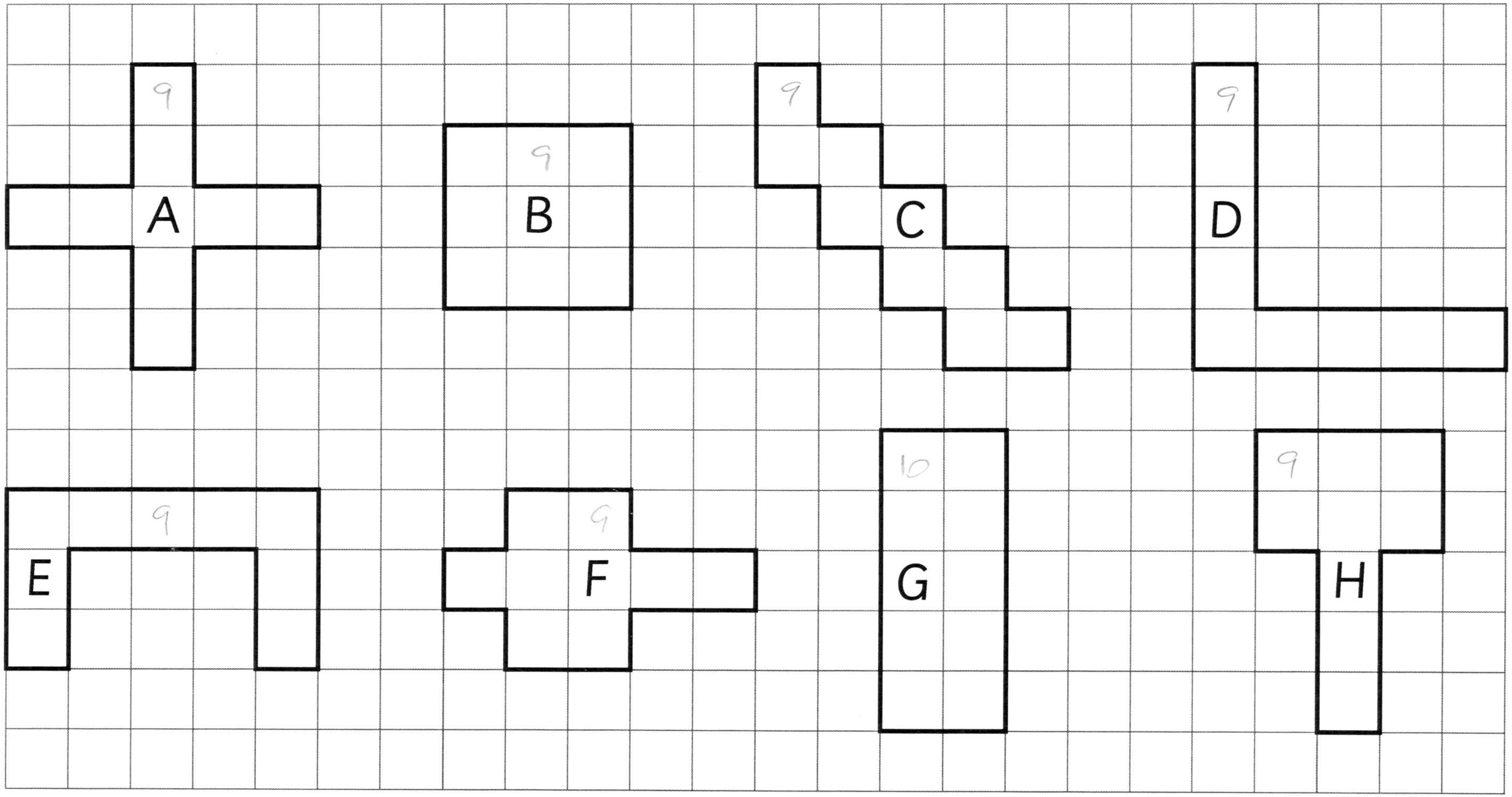

Mass 1

Learning objectives

● Choose and use standard metric units and their abbreviations when estimating, measuring and recording length, weight and capacity. (4Ml1)

● Know and use the relationships between familiar units of length, mass and capacity; know the meaning of kilo-, centi-, and milli-. (4Ml2)

Resources

Photocopiable page 43; a selection of everyday objects to find the mass of (e.g. fruit, packaged food items, classroom equipment); a range of standard masses (10 g, 100 g, 1 kg); a range of mass measuring instruments (pan balances, spring scales, kitchen scales and bathroom scales); an apple.

Starter

• Before the lesson make sets of cards from photocopiable page 43.

• Organise the learners into groups of between two and four. Give each group a set of cards.

• Ask the learners to match each mass in grams to the equivalent mass in kilograms, and then order the matched masses from lightest to heaviest.

• Challenge the learners to match the cards within a time limit, or challenge groups to compete with each other to be the first to finish.

Main activities

• Display a selection of everyday objects, a range of standard masses and a variety of instruments for measuring mass.

• Draw the following table on the board:

Object	Estimated mass	Measured mass	Difference

• Ask a volunteer to choose an object and estimate its mass. Encourage them to estimate by using direct comparison with the standard masses. Record the estimate in the table using an abbreviation (g or kg).

• Ask another volunteer to choose an instrument to measure the object's mass, and discuss reasons for that choice. Model using the instrument to measure the mass. Record the measurement using an abbreviation.

• Ask the learners to calculate the difference between the estimate and the measurement. Write the difference in the table.

• Repeat the process for an object with a very different mass from the first (so that you need to use a different measuring instrument).

• Organise the learners into pairs and ask them to copy the table then estimate and measure the mass of a variety of objects.

Plenary

• Ask the learners to look at the differences between their estimates and the actual measurements. Discuss whether the mass of some objects was easier to estimate than others and if so, why.

• Discuss whether estimates improved. Ask: *Did your estimates get better as you went along? Why do you think this is?*

Success criteria

Ask the learners:

● Can you write $2\frac{1}{2}$ kg in grams?

● Can you write 1250 g in kilograms?

● Estimate the mass of this apple. How did you work out your estimate?

● Measure the mass of the apple. How close was your estimate?

Ideas for differentiation

Support: When organising pairs for the final Main activity, group these learners with more confident learners who work well with others.

Extension: Challenge these learners to work on their own in the final Main activity.

Equivalent mass cards

100 g	500 g	1250 g
$\frac{1}{10}$ kg	$\frac{1}{2}$ kg	$1\frac{1}{4}$ kg
200 g	750 g	1500 g
$\frac{1}{5}$ kg	$\frac{3}{4}$ kg	$1\frac{1}{2}$ kg
250 g	1000 g	2000 g
$\frac{1}{4}$ kg	1 kg	2 kg

Capacity 1

Learning objectives

- Choose and use standard metric units and their abbreviations when estimating, measuring and recording length, weight and capacity. (4Ml1)
- Interpret intervals / divisions on partially numbered scales; record readings accurately. (4Ml4)
- Estimate and approximate when calculating, and check working. (4Pt8)
- Explain methods and reasoning orally and in writing; make hypotheses and test them out. (4Ps9)

Resources

Photocopiable page 45; a variety of waterproof containers (jugs, bottles, bowls, jars, mugs, cups, spoons, ladles, buckets); a variety of measuring jugs / cups / bowls marked in millilitres; water.

Starter

- Before the lesson, make a large set of cards from photocopiable page 45.
- Hold up the cards one at a time. Ask the learners to write the volume shown on each card, writing the units as abbreviations (l or ml). For volumes over 1 litre, discuss alternative ways of writing them (for example 1500 ml / 1 l 500 ml / $1\frac{1}{2}$ l / 1.5 l).

Main activities

- Hold up two containers, one with a capacity between two and ten times greater than the other (for example a tea cup and a jug). Ask: *About how many times could you fill the [smaller container] from the [larger container]?*
- Give the learners in pairs time to discuss their ideas. Agree on an estimate (for example 'about five times').
- Ask a volunteer to check the estimate by filling the larger container with water and repeatedly filling the smaller container with it. Record their measurement (for example 'a bit more than six times').

- Ask the learners to estimate the capacity of the smaller container, explaining their reasoning. Record an agreed estimate.
- Ask them to estimate the capacity of the larger container, explaining their method (for example multiplying the smaller container's estimated capacity by the number of times it can be filled by the larger container). Record an agreed estimate.
- Check the estimates by measuring the containers' actual capacities. Remind the learners to place the containers on a flat, level surface, and make sure their eyes are level with the top of the water.
- Organise the learners into groups, asking them to repeat the activity using pairs of containers of their own choice.

Plenary

- Ask the learners to look at the differences between their estimates and the actual measurements. Discuss whether the capacity of some containers was easier to estimate than others and if so, why.
- Discuss whether estimates improved. Ask: *Did your estimates get better as you went along? Why do you think this is?*

Success criteria

Ask the learners:

- Can you write $1\frac{1}{4}$ l in millilitres?
- Can you write 2000 ml in litres?
- Estimate the capacity of this cup. How did you work out your estimate?
- Measure the capacity of the cup. How close was your estimate?

Ideas for differentiation

Support: When organising pairs for the final Main activity, group these learners with more confident learners who work well with others.

Extension: Challenge these learners to devise strategies for measuring the capacity of very small or very large containers.

Reading volumes

Time 1

Learning objectives

- Read and tell the time to the nearest minute on 12-hour digital and analogue clocks. (4Mt1)
- Use am, pm and 12-hour digital clock notation. (4Mt2)

Resources

One large and lots of small clock faces with movable hands; photocopiable pages 47 and 48.

Starter

- Write the following times on the board: 3:45p.m., 2:45p.m., 2:45a.m. Ask the learners: *What time is a quarter to three in the afternoon?*
- Write the following times on the board: 7:40a.m., 8:20a.m., 8:40a.m. Ask: *What time is twenty to eight in the morning?*
- Write the following times on the board: 11:50a.m., 12:50a.m., 11:50p.m. Ask: *What time is ten minutes to midnight?* Ask them to read the other times.
- Call out a time to the nearest five minutes, for example: *half past midday / a quarter past six in the evening / twenty-five to eleven in the morning.* Ask the learners to write it as a digital time.

Main activities

- Before the lesson, make sets of cards from photocopiable pages 47 and 48.
- Display a time to the nearest minute on a large clock face. Ask the learners to say the time, explaining how they worked it out (for example by telling the time to the previous / next five-minute interval, and then adding / subtracting minutes).
- Hand out the small clock faces. On the board, write a digital time to the nearest minute. Ask the learners to make the time on their clock face.

- Organise the learners into groups of four. Give each group a set of cards made from photocopiable pages 47 and 48. Ask the learners to deal six cards to each player and put the rest in a face-down pile. A player starts their turn by drawing a card from the pile. If they have three cards in their hand showing the same time, they should place them face up on the table. At the end of their turn they should discard a face-up card. Subsequent players may choose to draw a card from the face-down pile or from the discard pile. The winner is the first player to get rid of all their cards.

Plenary

- Write a digital time on the board, to the nearest minute and more than 30 minutes past the hour. Ask the learners to show this time on an analogue clock face. Repeat.
- Make a time on an analogue clock face. Ask the learners to write this time in as many different ways as they can. Repeat.

Success criteria

Ask the learners:

- Is midday 12p.m. or 12a.m.?
- What is the time now?
- How many different ways can you write the time now?
- What will the time be in four hours' time? Can you show this time on a clock face?

Ideas for differentiation

Support: In the final Main activity, give these learners sets of cards from which the times written in words have been removed. Ask them to deal four cards to each player and collect pairs of cards instead of threes.

Extension: Ask these learners to devise their own games using the time cards.

Time cards 1

3.24p.m.	twenty-four minutes past three in the afternoon	3.24p.m.
3.24a.m.	twenty-four minutes past three in the morning	3.24a.m.
2.36p.m.	twenty-four minutes to three in the afternoon	2.36p.m.
2.36a.m.	twenty-four minutes to three in the morning	2.36a.m.
8.17p.m.	seventeen minutes past eight in the evening	8.17p.m.
8.17a.m.	seventeen minutes past eight in the morning	8.17a.m.

Time cards 2

7.43p.m.	seventeen minutes to eight in the evening	7.43p.m.
7.43a.m.	seventeen minutes to eight in the morning	7.43a.m.
1.28p.m.	twenty-eight minutes past one in the afternoon	1.28p.m.
1.28a.m.	twenty-eight minutes past one in the morning	1.28a.m.
12.32p.m.	twenty-eight minutes to one in the afternoon	12.32a.m.
12.32a.m.	twenty-eight minutes to one in the morning	12.32p.m.

Cambridge Primary: Ready to Go Lessons for Maths Stage 4 © Hodder & Stoughton Ltd 2013

Time 2

Learning objectives

- Understand everyday systems of measurement in length, weight, capacity and time, and use these to solve simple problems as appropriate. (4Pt2)
- Make up a number story for a calculation, including in the context of measures. (4Ps1)

Resources

Photocopiable page 50; clock faces with movable hands; calendars.

Starter

- Brainstorm units of time known by the learners, and write them on the board.
- Ask the learners to help you order the units of time from shortest to longest, and to define each unit of time in relation to at least one other unit, for example one year is 365 / 366 days, or 12 months or about 52 weeks.

Main activities

- Display an enlarged copy of photocopiable page 50, and read out one of the word problems.
- Ask the learners to describe the calculation they will need to perform in order to solve the problem, for example to solve the problem in Question 1, you need to calculate the time 1 hour and 25 minutes before 9p.m.
- Ask the learners to estimate the answer to the problem (for example about half past seven) and describe the strategies they used (for example rounding to the nearest half hour and then counting back).
- Work through the problem, asking the learners to suggest the method. Use a clock face and a calendar to support calculations where appropriate.

- Ask the learners to compare the answer with the estimate. Ask: *Given the estimate, does the answer seem reasonable? What would it mean if the estimate and the answer were very different?* (It might mean the answer is wrong.) *What else could you do to check the answer?* (You could repeat the calculation using a different method.)
- Work through another word problem in the same way.
- Hand out photocopiable page 50, clock faces and calendars. Ask the learners to complete the photocopiable page, working either individually or in pairs. Remind them to make an estimate for each question and use it to check the reasonableness of their answer.

Plenary

- Ask the learners to give the answers to the word problems on the photocopiable page, and describe the methods they used.
- Ask volunteers to read out the word problems they wrote and ask the rest of the class to solve one of the problems.

Success criteria

Ask the learners:

- How many days are there in one year?
- How many minutes are there in two hours?
- Choose one of the problems on photocopiable page 50 that you have already answered. How did you work out the answer?
- Can you make up a word problem about an event that starts at 4p.m. and lasts for one and a half hours?

Ideas for differentiation

Support: Group these learners together and guide them through an extra problem on photocopiable page 50. Ask them to miss out questions 5 to 7.

Extension: Ask these learners to make up word problems based on their own ideas instead of the questions given on photocopiable page 50.

Name: _____

Time problems 1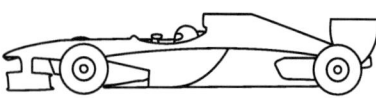

1. Sam and Ben want to watch a film on DVD that is 1 hour 25 minutes long.
    Afterwards, they want to watch a TV programme that starts at 9:00p.m.

    What is the latest time they should start watching the film?_____

    _____

2. The dishwasher cycle takes 1 hour 40 minutes. If I start the

    dishwasher at 7:25p.m., at what time will the cycle be finished?_____

    _____

3. Kristina started an 18-month mobile phone contract in April 2010.

    In which month and year did the contract end?

    _____

4. In how many years' time will it be 2050?_____

    _____

5. Sherif celebrated his 21st birthday in 2010. In what year was he born?

    _____

6. A racing car recorded these times in a race:
    Lap 1: 1 minute 35 seconds   Lap 2: 1 minute 30 seconds
    Lap 3: 1 minute 25 seconds   Lap 4: 1 minute 20 seconds.

    What was the car's total time for the race? _____

    _____

7. Write a word problem to go with each of these time phrases.
    Use the back of this page if you need more space:

    a)   three and a half hours before 2:30p.m. _____

    _____

    b)   three weeks later than 21 July _____

    _____

    c)   eighteen years before 2012 _____

    _____

          Cambridge Primary: Ready to Go Lessons for Maths Stage 4 © Hodder & Stoughton Ltd 2013

Time 3

Learning objectives

- Read simple timetables and use a calendar. (4Mt3)
- Understand everyday systems of measurement in length, weight, capacity and time, and use these to solve simple problems as appropriate. (4Pt2)
- Make up a number story for a calculation, including in the context of measures. (4Ps1)

Resources

Calendars; copies of the class timetable; photocopiable page 52.

Starter

- Organise the learners into pairs, giving each pair a calendar covering 12 months.
- Present the learners with various number calculations, the numbers for which can be found only by referring to the calendar, for example:
  - *Write down the date of the third Thursday in January and of the first Monday in November. Find the product of these two numbers.*
  - *Write down the date of the second Saturday in May, the third Wednesday in December and the second Sunday in April. Find the total of these three numbers.*

Main activities

- Ask the learners, still working in pairs, to use the calendar to work out the answers to problems involving finding durations of time in weeks and / or days, for example the number of days until the end of term, the number of weeks and days between two learners' birthdays, the length of time that has elapsed since a given event in the past, or the length of time until a given event in the future.

- Give each pair of learners a copy of the class timetable. Ask them to use it to work out the answers to various problems, for example how much time they spend: doing Maths / at break / in assembly / at lunch / in school every day and every week.
- Display an enlarged copy of photocopiable page 52 and read through the problems together.
- Hand out photocopiable page 52, asking the learners to work either independently or in pairs to complete it.

Plenary

- Ask the learners to share the answers to the word problems on photocopiable page 52, and explain how they worked them out.
- Ask volunteers to read out the word problems they wrote. Ask the other learners to solve one of the problems.

Success criteria

Ask the learners:

- How many days is it until the end of the year?
- How long do you spend each week in assembly?
- Choose one of the problems on photocopiable page 52 that you have already answered. How did you work out the answer?
- Can you make up another question about the TV schedule on photocopiable page 52?

Ideas for differentiation

Support: Group these learners together and work through one or two of the problems with them. Ask them to miss out questions 6 to 8.

Extension: Ask these learners to draw their own timetable on which to base the questions they write themselves.

Name: _____

TV schedule problems

Channel: Children's TV	
Time	Programme
3:30p.m.	Lambs' Tails
3:40p.m.	Space Frontiers
3:55p.m.	Animal Emergency
4:20p.m.	History Mysteries
4:40p.m.	Cooking the Books
5:30p.m.	Funny Bone
5:55p.m.	Amazing Adventures

Handwritten notes beside table: 10, 15, 25, 20, 50, 25, -6:20

1. Which is the shortest programme? How long is it? _____

2. How long is an episode of Animal Emergency? _____

3. Sereena watches Cooking the Books and Funny Bone.

    How long does she watch TV for? _____

4. It's a quarter to four when Paulo gets home from school.

    How long does he have to wait until he can watch History Mysteries?

    _____

5. Auzma gets home at 4:05p.m. What programme is on TV then?

    _____

    How many minutes is it until the next programme starts? _____

6. How much longer is Cooking the Books than Space Frontiers? _____

7. Arjun watches Lambs' Tails, Animal Emergency and Cooking the Books.

    He turns the TV off in between each programme.

    How long does Arjun watch TV for? _____

8. Amazing Adventures is 25 minutes long.

    How long do the seven programmes last altogether? _____

9. Make up three questions of your own that can be answered by

    reading the TV schedule. Write them on the back of this page.

Cambridge Primary: Ready to Go Lessons for Maths Stage 4 © Hodder & Stoughton Ltd 2013

Unit assessment

Questions to ask

- Name three units that are used to measure time. What is the relationship between them?
- What do the prefixes 'milli-', 'centi-' and 'kilo-' mean?
- What time is shown on this digital clock? Can you make the same time on an analogue clock face?

- Imagine a rectangle that is 6 cm long and 5 cm wide. What is its area? What is its perimeter? How did you work out your answers?
- Estimate the capacity of this container. How did you reach your estimate?
- What is $1\frac{3}{4}$ kilograms in grams?

Summative assessment activities

Observe the learners while they take part in these activities. You will quickly be able to identify those who appear to be confident and those who may need additional support.

Time match game

This game assesses the learners' ability to read and tell the time to the nearest minute on 12-hour digital and analogue clocks (including using a.m. and p.m.).

You will need:

Sets of 36 cards made from photocopiable pages 47 and 48.

What to do

- Organise the learners into groups of six. Give each group a set of 36 cards. Ask the dealer to shuffle the cards and deal out six cards to each player.
- The player to the left of the dealer must choose a card from their hand and put it face up on the table. The players with the two matching cards in the set of three should place them face up next to the other card. When everyone has checked that all three cards match, they should turn them face down. Play passes to the next player to the left.
- The game is over when one player has no more cards left in their hand. This player is the winner.

Same perimeter

This activity assesses the learners' ability to find the area of rectilinear shapes drawn on a square grid by counting squares, and their ability to measure and calculate the perimeters of rectangles.

You will need:

Squared paper; coloured pencils; rulers.

What to do

- Hand out the squared paper, coloured pencils and rulers.
- Ask: *Using a square grid, how many different rectangles can you make with a perimeter of 24 cm? Can you find all of them?* (There are six rectangles: 1 by 11, 2 by 10, 3 by 9, 4 by 8, 5 by 7 and 6 by 6.)
- Ask the learners to find the area of each rectangle. Ask: *Which of the shapes has the largest / smallest area?* (The 6 by 6 rectangle – the square – has the largest area, and the 1 by 11 rectangle has the smallest area.)

Written assessment

Hand out photocopiable page 54. Ask the learners to read the questions and write the answers. They should work independently.

Measure problems

1. What is the mass of the pineapple? _____

2. Choose an object and write it here: _____

   a) Estimate its mass: _____

   b) Measure its mass: _____

   c) Calculate the difference between your estimate and the measurement.

   Rgl

   _____

   _____

3. What units would you use to measure the length of time until your next birthday?

   _____

4. A bus was delayed by 38 minutes. It was supposed to depart at 7:45a.m.

   What time did it actually depart? _____

5. Choose a waterproof container and write it here: _____

   a) Estimate its capacity: _____

   b) Measure its capacity: _____

   c) Calculate the difference between your estimate and the measurement.

   _____

6. How many millilitres are there in two and a half litres? _____

7. Write a question that can be answered by reading this timetable.

Time	8:30– 8:45	8:45– 9:15	9:15– 10:15	10:15– 10:40	10:40– 11:40	11:40– 12:10	12:10– 12:50	12:50– 1:15	1:15– 2:15	2:15– 2:30	2:30– 3:15
Mon	Register	Assembly	Maths	Morning break	Literacy	ICT	Lunch	Reading	Art	Afternoon break	PE

_____

_____

Now answer your question. _____

Cambridge Primary: Ready to Go Lessons for Maths Stage 4 © Hodder & Stoughton Ltd 2013

Handling data 1

55

Learning objectives

● Answer a question by identifying what data to collect, organising, presenting and interpreting data in tables, diagrams, tally charts, frequency tables, pictograms and bar charts. (4Dh1)

● Use ordered lists and tables to help to solve problems systematically. (4Ps5)

● Explain methods and reasoning orally and in writing; make hypotheses and test them out. (4Ps9)

Resources

Photocopiable pages 56 and 57; squared paper; rulers; coloured pencils.

Starter

• Before the lesson, make sets of cards from photocopiable pages 56 and 57.

• Organise the learners into pairs and give each pair a set of cards. Ask them to match each diagram with its name.

• Confirm the name of each diagram, and then ask the learners questions requiring them to interpret the data in each diagram, for example:
  • *Look at the bar chart. Which are the two most popular colours?*
  • *Look at the pictogram. How many more birds were seen in the playground on Friday than on Monday?*

Main activities

• Ask the learners in pairs to devise a question that can be answered by collecting data. They could use the diagrams from the Starter activity as inspiration (for example: 'What is the most popular sport in our school?'). Alternatively, they could link their question to current themes in other areas of the curriculum.

• Take the learners' suggestions and write on the board any that are practicable. Ask the learners to predict the answer to each question and explain their reasoning. Ask them what data they will need to collect to answer the question, and how they might collect it.

• Ask the learners to choose one of the questions on the board. Group the learners according to their choice of question. (Several small groups working on a question works better than one large group.) Ask the learners to collect the data they need, and present it in the form of a diagram of their choice.

Plenary

• Ask a volunteer from each group to describe which question they chose, what data they collected and how they collected it. Ask them to show the diagram their group drew.

• Ask each group whether they have enough information to answer the question they chose. If they do, ask them to give the answer. If they don't, ask them to suggest what extra data they could collect that might allow them to answer the question.

Success criteria

Ask the learners:

● What question are you trying to answer?

● What do you think the answer to the question will be? Why?

● What data are you collecting? How will this data help you to answer the question you're trying to answer?

● What does the data in this diagram tell you?

Ideas for differentiation

Support: To support these learners, organise mixed-ability groupings in the final Main activity.

Extension: In the final Main activity, ask these learners to draw a second diagram showing the same data as the first, and to explore the differences between the two diagrams.

Data handling cards 1

tally chart

Sport	Tally
Athletics	卌 卌 I
Basketball	卌 卌 II
Cricket	卌 卌 卌 IIII
Cycling	卌 III
Football	卌 卌 卌 卌 卌 I
Netball	卌 I
Rugby	III
Swimming	卌 卌 卌 卌 I

frequency table

Sport	Frequency
Football	26
Swimming	21
Cricket	19
Basketball	12
Athletics	11
Cycling	8
Netball	6
Rugby	3

bar chart

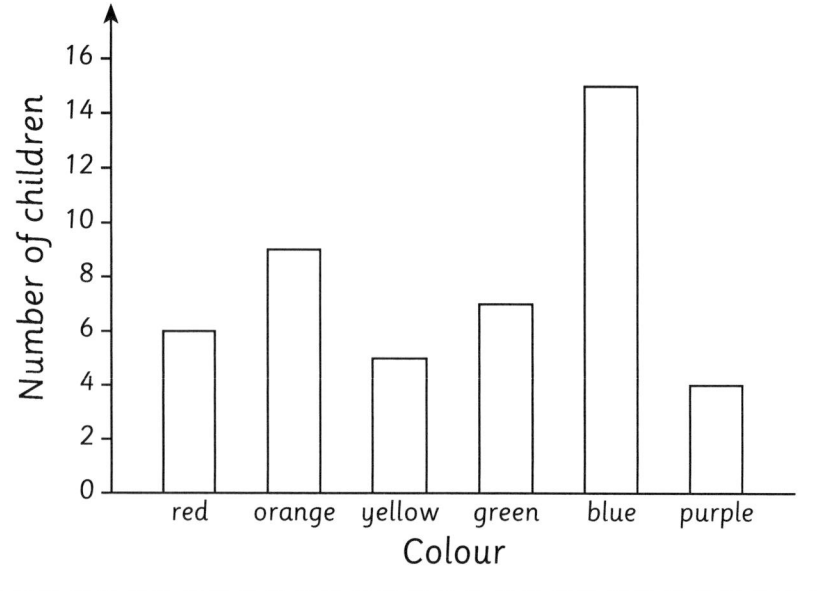

Data handling cards 2

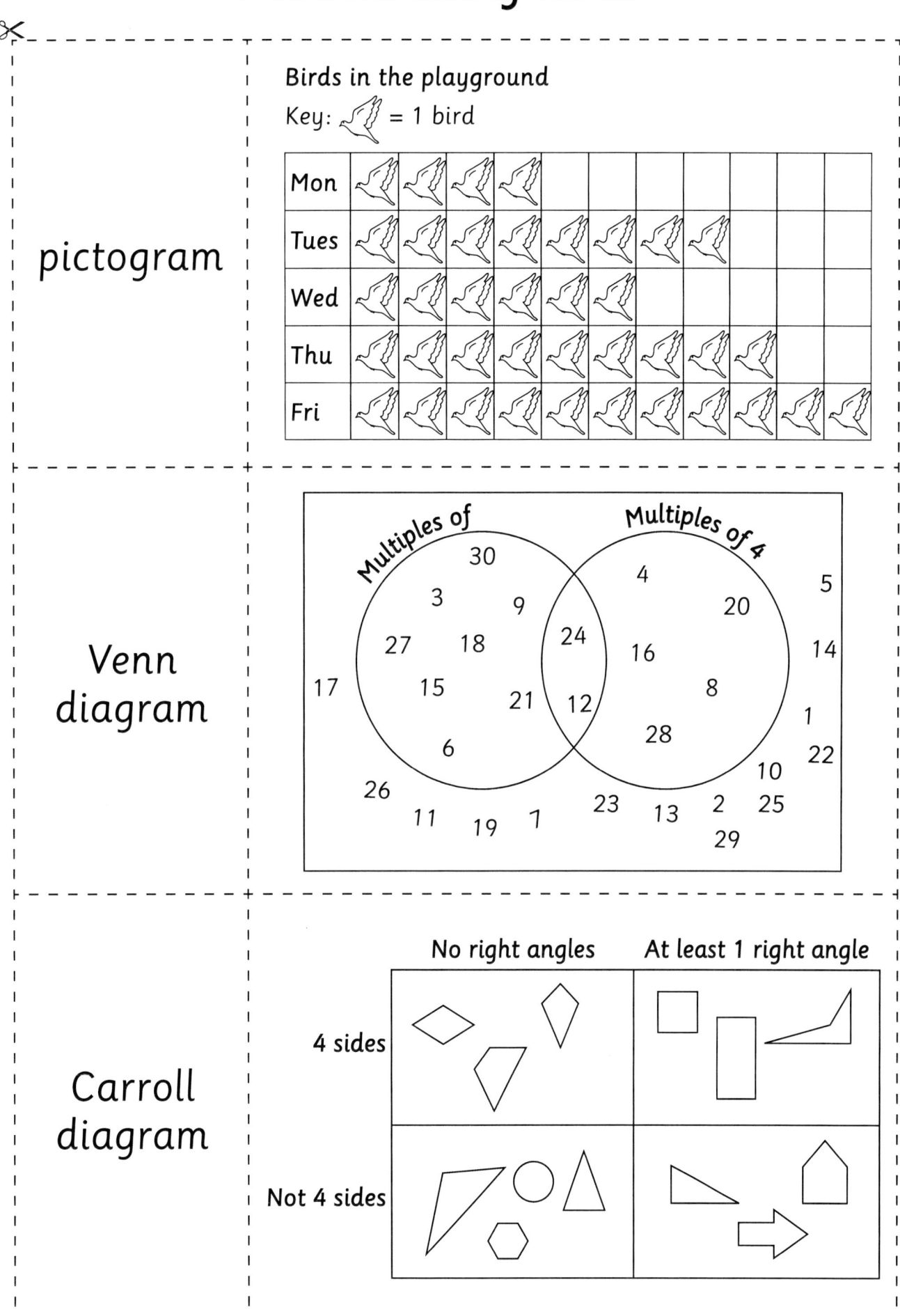

pictogram

Birds in the playground

Key: = 1 bird

Mon	🕊	🕊	🕊	🕊						
Tues	🕊	🕊	🕊	🕊	🕊	🕊	🕊			
Wed	🕊	🕊	🕊	🕊						
Thu	🕊	🕊	🕊	🕊	🕊	🕊	🕊			
Fri	🕊	🕊	🕊	🕊	🕊	🕊	🕊	🕊	🕊	🕊

Venn diagram

Multiples of   Multiples of 4

30
3   9   4   5
27   18   24   20
17   15   16   14
21 12   8
6   28   1
26   22
11   19   7   23   13   2   25   10
29

Carroll diagram

	No right angles	At least 1 right angle
4 sides		
Not 4 sides		

Handling data 2

Learning objectives

● Answer a question by identifying what data to collect, organising, presenting and interpreting data in tables, diagrams, tally charts, frequency tables, pictograms and bar charts. (4Dh1)

● Compare the impact of representations where scales have different intervals. (4Dh2)

Resources

Counting stick; photocopiable page 59; squared paper; rulers; coloured pencils.

Starter

- Using the counting stick, lead the class in counting on from 0 in 2s and back again.
- Point to each division on the counting stick in a random order, asking the learners to say each number aloud, for example if you point to the seventh division, the learners should say 'fourteen'.
- Repeat the activity for counting in 5s, 10s and 20s.

Main activities

- Display a copy of Bar chart 1 from photocopiable page 59. Ask an open-ended question, for example: *What does this bar chart tell you?*
- Introduce the terms 'horizontal axis' and 'vertical axis', and mark these on Bar chart 1. Ask the learners to describe the scale on the vertical axis. (It starts at 0 and counts in 10s.)
- Display a copy of Bar chart 2 from photocopiable page 59, alongside Bar chart 1. Ask: *Do these two bar charts show the same data? How can you tell?*
- Ask the learners to describe the scale on the vertical axis in Bar chart 2. (It starts at 20 and counts in 5s.) Ask what difference this makes to the way the data looks. Ask: *Which chart would you choose if you wanted to suggest sales do not vary much much from week to week? Why?*

- Distribute photocopiable page 59. Ask the learners to draw two more versions of the bar chart, one whose vertical axis starts at 20 and counts in 2s and one whose vertical axis starts at 0 and counts in 20s.

Plenary

- Ask the learners to compare all four versions of the bar chart, and describe the differences and similarities between them.
- Ask them to explain which chart they would choose if they wanted to emphasise how much sales vary from week to week, and why.

Success criteria

Ask the learners:

● How does using a different scale on the vertical axis change the way a bar chart looks?

● Imagine drawing a bar chart showing the same data, but whose vertical scale started at 0 and each division represented 1. What would the bars look like on that bar chart?

● Imagine drawing a bar chart showing the same data, but whose vertical scale started at 0 and each division represented 50. What would the bars look like on that bar chart?

● Why might you want to choose the scale carefully when drawing a bar chart?

Ideas for differentiation

Support: In the final Main activity, group these learners together. Ask them to draw just one of the bar charts each, and help them to label the vertical axis where necessary.

Extension: Ask these learners to draw a third bar chart using a different scale of their choice.

Comparing scales

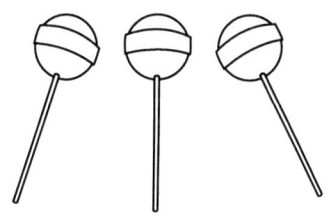

Bar chart 1

Sales at the sweet shop

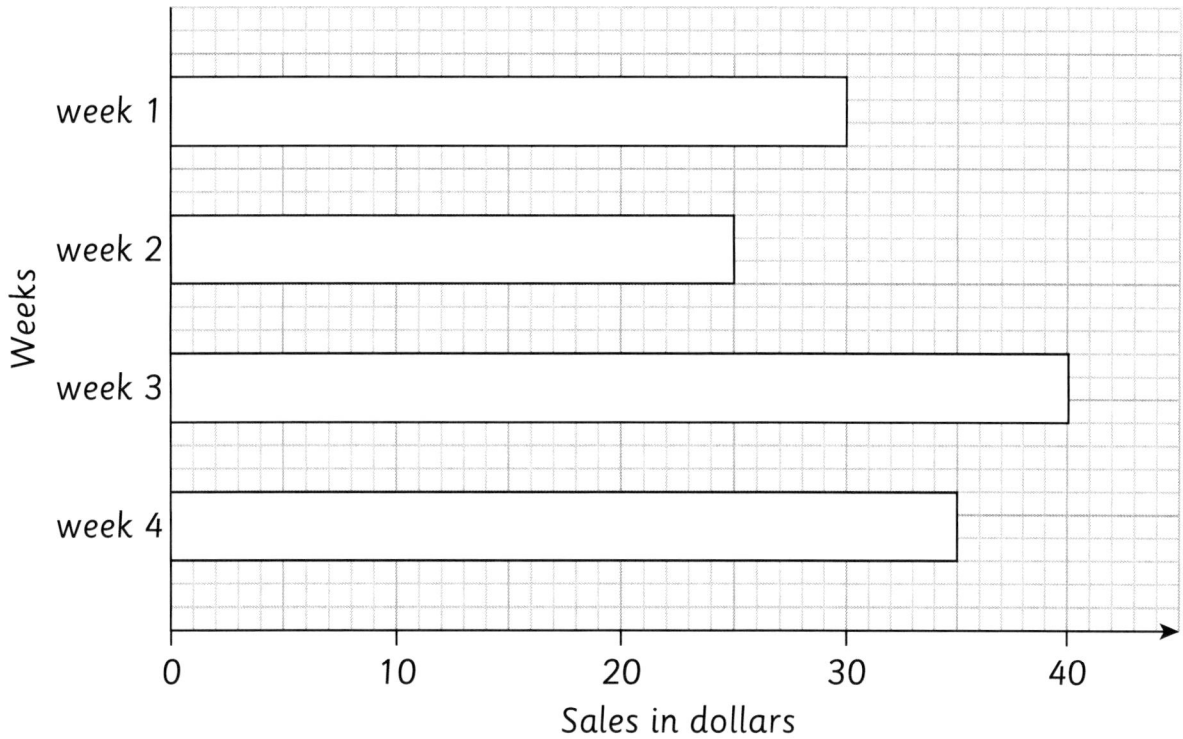

Sales in dollars

Weeks: week 1, week 2, week 3, week 4

✂ —

Bar chart 2

Sales at the sweet shop

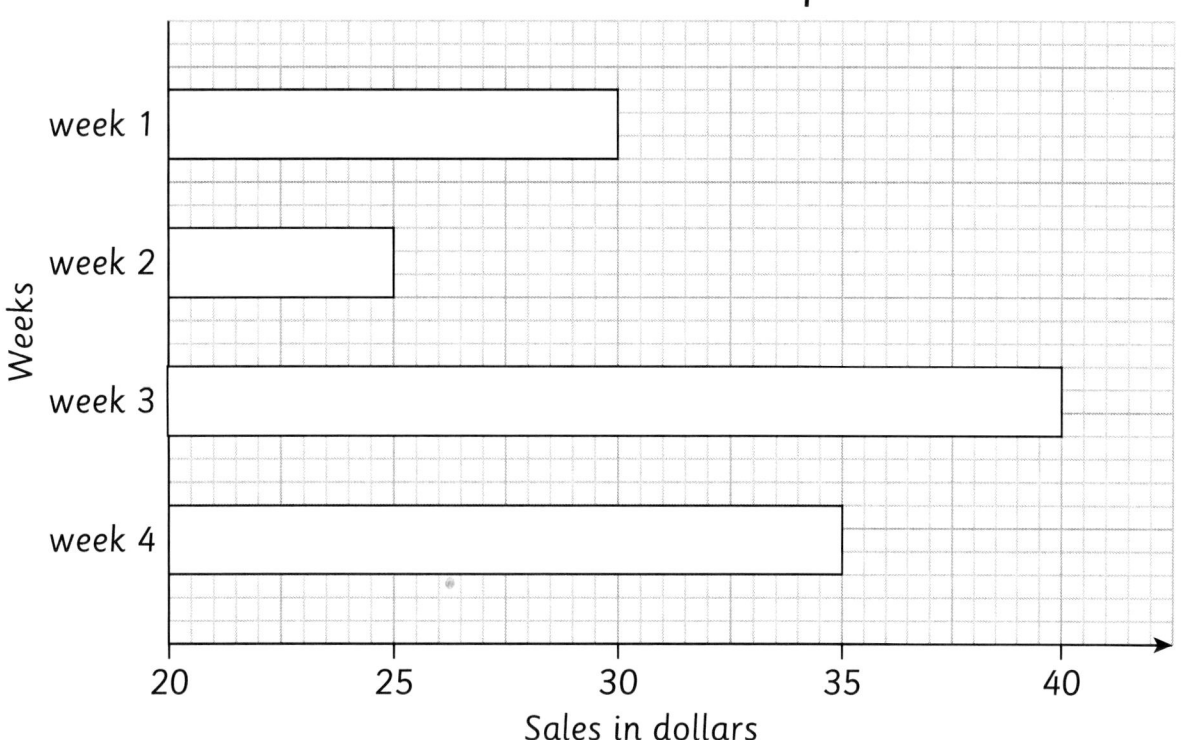

Sales in dollars

Weeks: week 1, week 2, week 3, week 4

Handling data 3

Learning objectives

● Use Venn or Carroll diagrams to sort data and objects using two or three criteria. (4Dh3)

Resources

Photocopiable page 61; A3 paper; 2D shapes.

Starter

• Draw a two-circle Venn diagram on the board. Label one circle 'even' and the other 'multiples of 3'.

• Ask the learners to copy the diagram into their exercise books and write each whole number from 0 to 15 in the correct region of the diagram. Share answers.

• Repeat the activity, relabelling the circles 'less than 4' and 'factors of 12'. Share answers.

Main activities

• Display the Venn diagram on photocopiable page 61. Ask the learners questions about the data, for example:
  • *Which animals in the diagram have wings and six legs?* (Butterflies and bees.)
  • *Which animals in the diagram have neither wings nor six legs?* (Centipedes, worms, woodlice, spiders.)

• Draw a Venn diagram on the board. Label the circles 'born January to June' and 'even birth date'. Ask the learners to put their hands up when you point to 'their' region of the diagram. Record names or initials.

• Revise Carroll diagrams using the Carroll diagram on photocopiable page 61. Ask the learners questions about the data. Ask them to compare the Carroll diagram to the Venn diagram on photocopiable page 61, and establish that the diagrams show the same data.

• On the board, draw a Carroll diagram. Label the columns 'born January to June' and 'born July to December', and label the rows 'even birth date' and 'not even birth date'. Ask the learners to put their hands up when you point to 'their' region of the diagram. Record names or initials.

• Organise the learners into groups of six to eight and hand out A3 paper. Ask them to sort the group members according to their own criteria. Ask them to record the data in a Venn diagram and in a Carroll diagram.

Plenary

• Ask groups to share and explain the Venn and Carroll diagrams they have drawn.

• Draw a Venn diagram or a Carroll diagram on the board. Write in data, for example shapes or numbers, but do not write in labels for the circles / boxes. Ask the learners to suggest what the labels might be.

Success criteria

Ask the learners:

● Where does this shape / number belong in this Carroll diagram? Why?

● Where does this shape / number belong in this Venn diagram? Why?

● Can you draw a Venn diagram to sort these 2D shapes?

● Can you draw a Carroll diagram to sort these numbers?

Ideas for differentiation

Support: Group these learners together and help them to devise the sorting criteria they are going to use, ensuring that each pair of criteria in the Carroll diagram covers all possible answers.

Extension: Ask these learners to suggest other data / objects to sort using a Venn or Carroll diagram.

Name: _____

Venn and Carroll diagrams

A Venn diagram

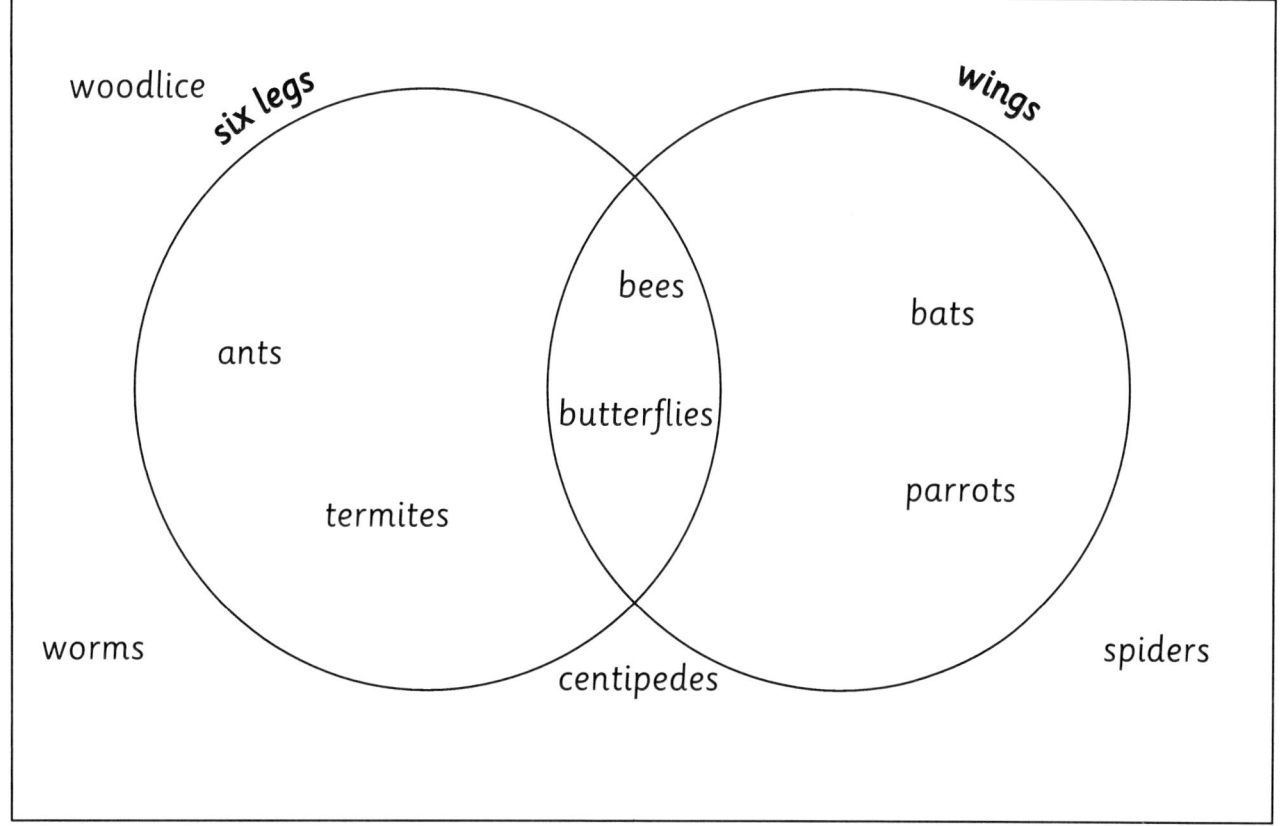

Animals we have studied

woodlice

six legs

wings

bees

bats

ants

butterflies

parrots

termites

worms

centipedes

spiders

✂ -

A Carroll diagram

Animals we have studied

	six legs	not six legs
wings	bees butterflies	bats parrots
no wings	ants termites	woodlice worms centipedes spiders

Handling data 4

Learning objectives

- Answer a question by identifying what data to collect, organising, presenting and interpreting data in tables, diagrams, tally charts, frequency tables, pictograms and bar charts. (4Dh1)
- Explain methods and reasoning orally and in writing; make hypotheses and test them out. (4Ps9)

Resources

Times table charts for 6 and 9 times tables; sticky notes; photocopiable page 63; potato chips in a variety of flavours; blindfolds; tape measures; metre sticks.

Starter

- Display a chart for the 6 times table and practise chanting it. Each time you chant, cover up a couple of products (answers) with sticky notes until all the products are covered up. Repeat for the 9 times table.

Main activities

- Display a copy of photocopiable page 63, which poses a variety of questions that can be answered by collecting numerical data. For each question, ask the learners to predict the answer and explain their reasoning.
- For each question, ask the learners what data they would need to collect in order to answer it, and to explain how they might collect it, for example: *What data could be collected by asking people?* (Questions 1, 2, 5 and 6.) *What data could be collected by taking measurements?* (Question 4.) *What data could be collected by doing an experiment and making observations?* (Question 3.)
- Ask: *How could you present the data after you have collected it?* Revise tally charts, frequency tables, bar charts and pictograms as ways of organising and presenting data. Ask the learners to suggest which type of chart or table might be suitable for presenting the data for each question, and explain their reasoning.

- Organise the learners into groups. Ask each group to choose one of the questions on photocopiable page 63 to answer. Explain that before they start collecting data they will need to decide which group of people they will collect data about, for example the group / class / year group / school. Ask each group to decide for themselves how they will collect the data and how they will present it.

Plenary

- Ask a representative from each group to say which question they chose, and which group of people they collected the data from.
- Ask them to describe how they collected the data, display the chart or graph they drew and explain what it shows.

Success criteria

Ask the learners:

- What question did you choose?
- What data did you collect and how did you collect it?
- How did you organise and present the data?
- What does the data you collected show?

Ideas for differentiation

Support: Organise all the learners into mixed-ability groups, so that these learners are supported by other group members.

Extension: Ask these learners to ask and answer their own follow-up question to the initial question they investigated.

Collecting data

Choose one of these questions to answer:

1. Which is the most popular school subject?

2. Do some months of the year have more birthdays in than others?

3. Can blindfolded people identify potato chip flavours?

4. Do people with large feet also have large hands?

5. What changes would children like to see in the playground?

6. What is the most common number of children in each family?

Unit assessment

- Display photocopiable page 65, and ask the following questions:
  - Do these two bar charts show the same data? How can you tell?
  - Which chart would you choose if you wanted to emphasise Tom's improvement in spelling? Why?

- Between which two consecutive weeks did Tom's score increase by the largest amount?
- Between which two consecutive weeks did Tom's score go down?
- By how many points did Tom's test score improve from Week 1 to Week 6?
- What other questions could you ask about the data in the bar charts?

Summative assessment activities

Observe the learners while they take part in these activities. You will quickly be able to identify those who appear to be confident and those who may need additional support.

Drawing a pictogram

This activity assesses the learners' ability to interpret frequency tables and draw pictograms.

You will need:

Squared paper; rulers; pencils; erasers; coloured pencils.

What to do

- Draw a frequency table on the board. You could use or adapt the table below.

| Number of balloons sold in one week at Bobbie's Balloons ||
Day	Number of balloons
Monday	10
Tuesday	25
Wednesday	12
Thursday	20
Friday	35
Saturday	58
Sunday	40

- Ask the learners to draw a pictogram of the data shown in the frequency table. Remind them that they will need to include a key, to show how many balloons each picture represents.

Sorting data

This activity assesses the learners' ability to organise data into three-circle Venn diagrams.

You will need:

Ten-sided dice (0 to 9) or 0 to 9 digit cards.

What to do

- Draw a Venn diagram on the board, with circles labelled 'even' and 'less than 500'. Ask the learners to copy the diagram.

- Give each learner a ten-sided dice or a set of 0 to 9 digit cards. Ask the learners to use the dice or cards to generate three-digit numbers, and then write each number in the correct region of the diagram.

- When the learners have recorded 20 numbers in the diagram, challenge them to repeat the activity using a new Venn diagram that they label with their own criteria.

Give each learner a copy of the Carroll diagram from photocopiable page 61. Ask them to write the names of two more animals in each region of the diagram.

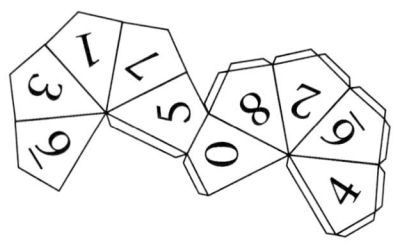

Bar charts

Bar chart 1

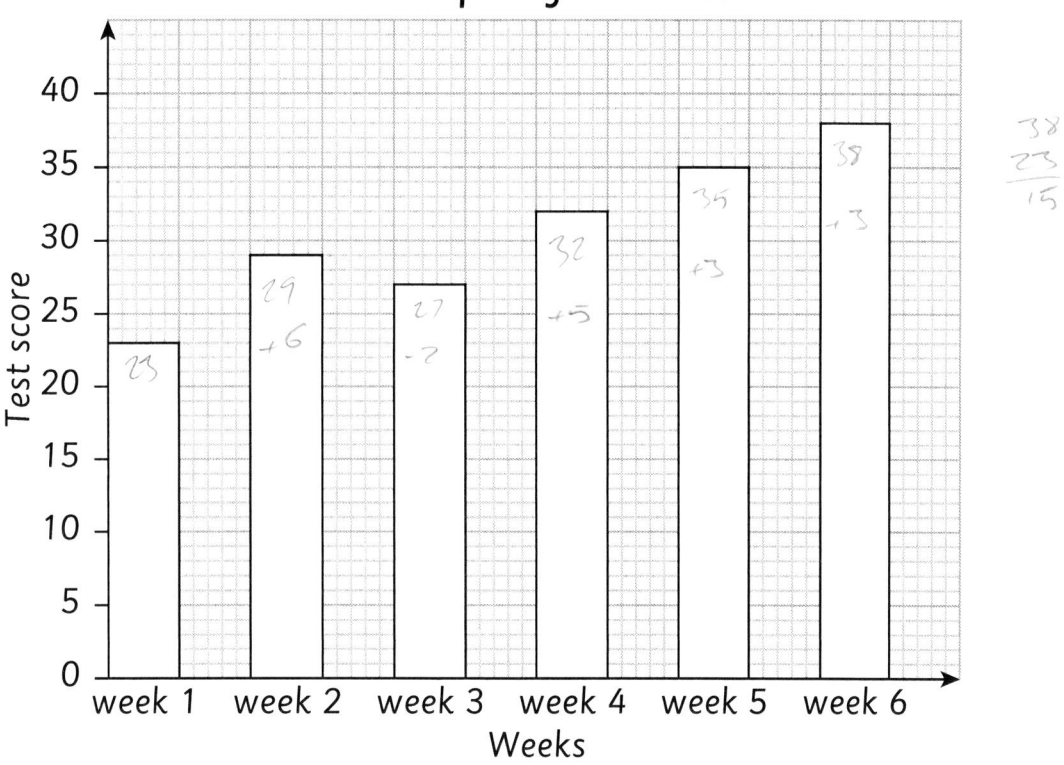

Tom's spelling test scores

Bar chart 2

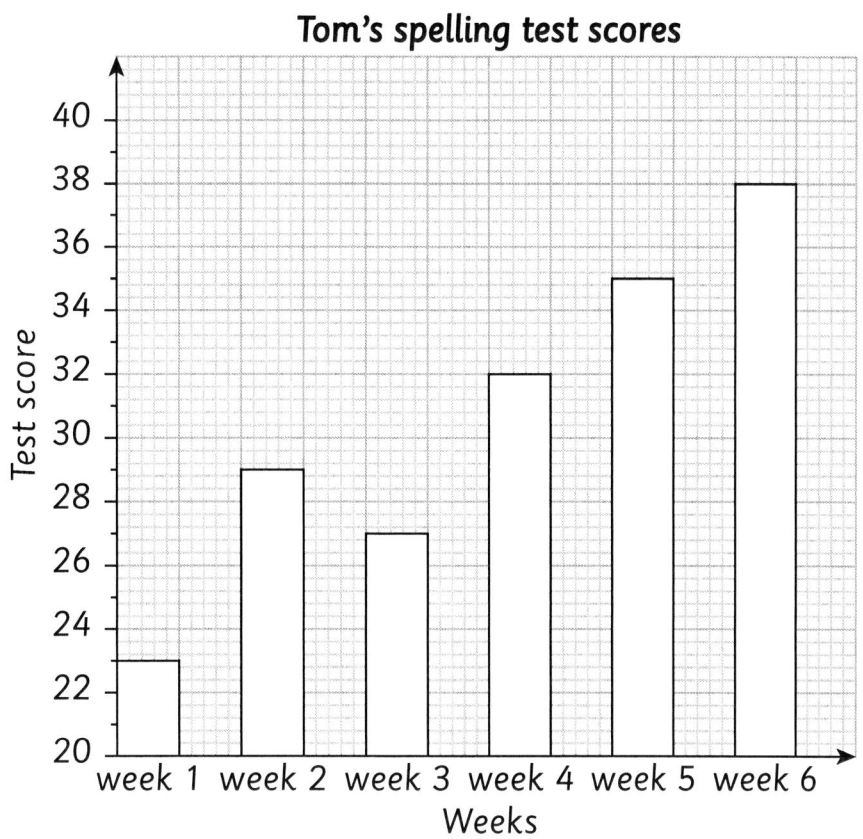

Tom's spelling test scores

Unit 2A: Number and problem solving

Place value 1

Learning objectives

- Understand what each digit represents in a three- or four-digit number and partition into thousands, hundreds, tens and units. (4Nn3)
- Use decimal notation and place value for tenths and hundredths in context. (4Nn4)
- Multiply and divide three-digit numbers by 10 (whole number answers) and understand the effect; begin to multiply numbers by 100 and perform related divisions. (4Nn7)
- Round three- and four-digit numbers to the nearest 10 or 100. (4Nn9)

Resources

Place value cards (Th, H, T, U); ten-sided dice; photocopiable page 67; sticky notes; calculators.

Starter

- Make a four-digit number from place value cards, for example 2673. Ask the learners to round it to the nearest 10 (2670) and 100 (2700). Discuss strategies. Ask: *Which cards make this number?* Pull the cards apart, writing: 2673 = 2000 + 600 + 70 + 3.
- Repeat for other four-digit numbers, including some containing 0s, or whose units or tens digit is a 5.

Main activities

- Display a large copy of photocopiable page 67. Write a three-digit number on a sticky note and put it in the 'in' box. Put a sticky note reading '×10' in the 'function' box. Ask the learners to find the output, explaining their method. Repeat.
- Put a sticky note with a four-digit multiple of 10 in the 'out' box, asking the learners to find the input. Ask: *When you found the output from the input, you were multiplying by 10. What were you doing when you found the input from the output?* (Dividing by 10.) Repeat.

- Replace ×10 on the function machine with ÷10. Ask the learners to predict what might happen to numbers put into the machine now.
- Organise the learners into pairs, giving each pair a calculator. Ask the learners to test their ideas by dividing two- and three-digit numbers by 10. Explain how to read numbers with a decimal point. Ask: *Can you find an input number that gives an output without a decimal point?*

Plenary

- Ask the learners to describe general rules for multiplying and dividing numbers by 10.
- Ask the learners which input numbers give an output without a decimal point. Ask: *What do you notice about those numbers?* (They are multiples of 10 / end in a 0.)

Success criteria

Ask the learners:

- What is 2439 rounded to the nearest 10 / 100?
- Make a four-digit number using the cards. Read the number aloud. What is each digit worth?
- What is 1340 divided by 10?
- Can you write a number that gives an answer without a decimal point when you divide it by 10?

Ideas for differentiation

Support: In the final Main activity, work with these learners using a calculator.

Extension: In the final Main activity, challenge these learners to start dividing numbers by 100 and 1000 and make a note of any patterns they see.

Function machine

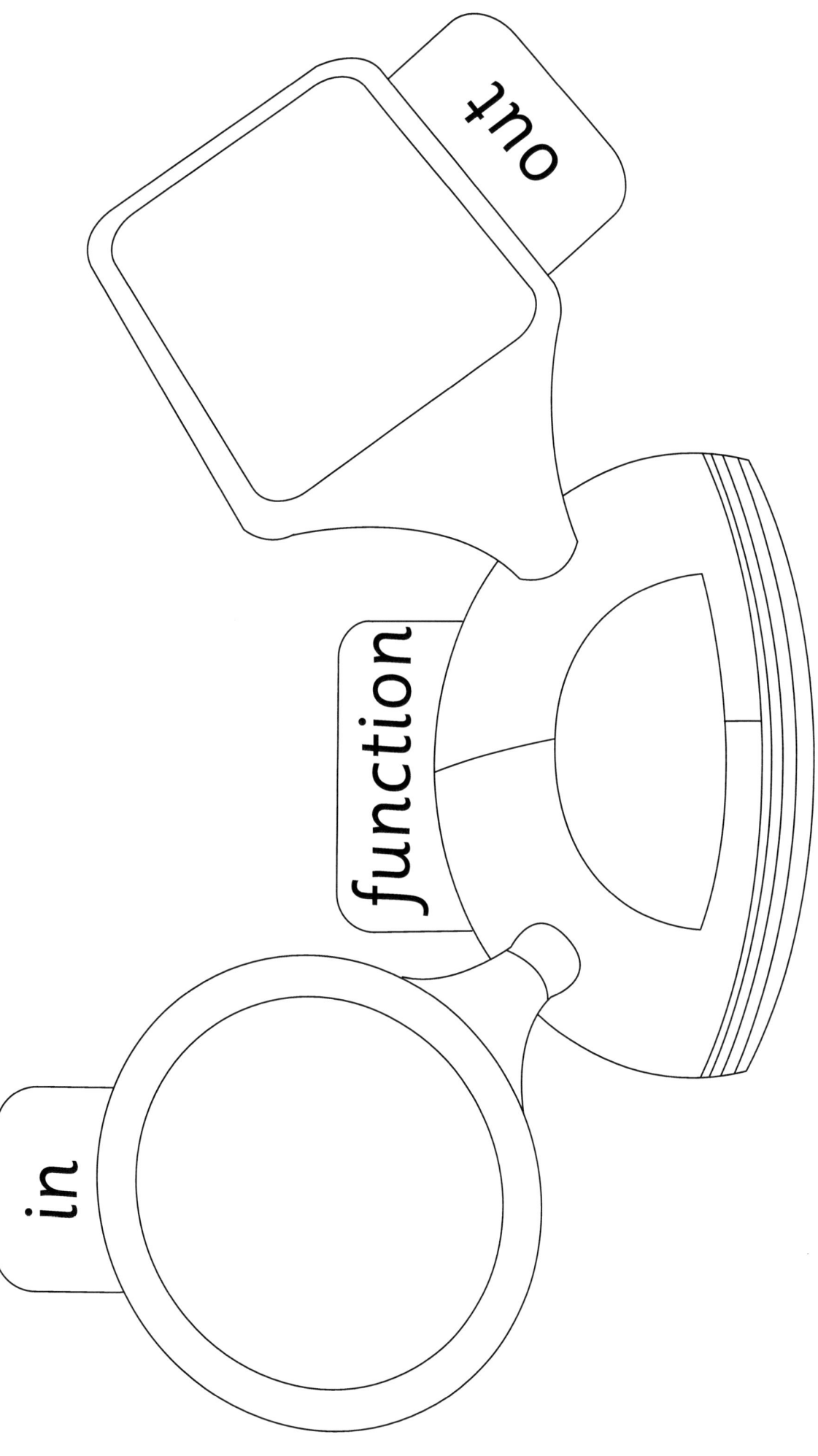

Place value 2

Learning objectives

● Count on and back in ones, tens, hundreds and thousands from four-digit numbers. (4Nn2)
● Find multiples of 10, 100, 1000 more / less than numbers of up to four digits, e.g. 3407 + 20 = 3427. (4Nn6)

Resources

Counting stick; large and small place value cards; cards made from photocopiable page 69; six-sided dice with three faces labelled 'less than' and three faces labelled 'more than'; 0 to 9 digit cards; calculators.

Starter

- Using the counting stick, lead the class in counting on and back from four-digit numbers in 1s, 10s, 100s and 1000s. Emphasise the changing digit in each number, for example: *six thousand five hundred and thirty-**three**, six thousand five hundred and thirty-**four** … / three thousand **six** hundred and eighteen, three thousand **seven** hundred and eighteen …*

Main activities

- Organise the learners into pairs, giving each pair a set of place value cards.
- Make a three- or four-digit number with the large place value cards. Ask the learners to find a multiple of 10 more or less and make the answer with their place value cards. Include some examples that involve bridging 100, for example 50 more than 8471 / 30 less than 2915.
- Repeat the activity, finding multiples of 100 more or less, including some examples bridging 1000. Finally, find multiples of 1000 more or less.
- Organise the pairs into groups of six or eight. Give each group a set of cards made from photocopiable page 69, a six-sided dice labelled 'less than' and 'more than', a set of 0 to 9 digit cards and a calculator.

- Ask the learners to use the digit cards to generate a four-digit number, and then find how they need to change this number by drawing a card from photocopiable page 69 (to give a multiple of 10, 100 or 1000) and then rolling the dice (to give 'more than' or 'less than').
- The pairs should make the new number on their place value cards. One learner should check the answer using the calculator, and every pair with the right answer scores a point.

Plenary

- Ask the learners some quick-fire questions that require them to find multiples of 10, 100 or 1000 more or less than numbers of up to four digits.
- Ask: *In the game, did you ever generate a question you couldn't answer? Give an example.* (Subtracting a multiple of 1000 from a number whose thousands value is less than the multiple of 1000). *What did you do?*

Success criteria

Ask the learners:

● Can you count on in hundreds from 5809?
● What is 70 less than 2159?
● What is 400 more than 4272?
● What is 3000 less than 9408?

Ideas for differentiation

Support: In the final Main activity, group these learners together and give them a set of cards made from photocopiable page 69 from which the multiples of 1000 have been removed.

Extension: Challenge these learners to do the group Main activity without using the place value cards.

Multiples of 10, 100 and 1000 cards

20	30	40	50
60	70	80	90
200	300	400	500
600	700	800	900
2000	3000	4000	5000
6000	7000	8000	9000

Odds and evens 1

Learning objectives

- Recognise odd and even numbers. (4Nn15)
- Make general statements about the sums and differences of odd and even numbers. (4Nn16)
- Investigate a simple general statement by finding examples which do or do not satisfy it. (4Ps8)

Resources

Cards made from photocopiable page 71; six-sided dice labelled with three + signs and three − signs.

Starter

- On the board write various numbers of between two and four digits.
- Ask the learners to sort the numbers into two groups: odd and even. Ask: *How do you know which number is odd and which is even?* (By looking at the units digit of each number.) Ask the learners to formulate a general rule. (A number whose units digit is 0, 2, 4, 6 or 8 is even; a number whose units digit is 1, 3, 5, 7 or 9 is odd.)

Main activities

- Ask the learners to choose two even numbers from the board and find their total. Ask: *Is the answer even or odd?* Ask them to repeat with another pair of even numbers. Ask: *What do you notice?* (The total of two even numbers is always even.)
- Ask the learners to investigate the statement: 'The difference between an even number and an odd number is always odd'.
- Ask the learners to make their own general statements about adding and subtracting pairs of odd and even numbers and test them out to see whether they are true.
- Collate all the tested statements.

- Organise the learners into pairs. Give each pair a set of cards made from photocopiable page 71 and a dice (see 'Resources'). Ask the learners to shuffle the cards and place them in a face-down pile. Decide which player is 'evens' and which is 'odds'. Turn over two cards (for example 420 and 675) and roll the dice (for example −) to generate a number sentence (for example 675 − 420). Ask the learners to predict if the answer will be odd or even using their knowledge from the previous activity, before working out the answer. If the answer is even, the 'evens' player wins the cards. If it is odd, the 'odds' player wins them. The winner is the player with the most cards when there are no cards left in the draw pile.

Plenary

- Discuss the game with the learners. Ask them which player won the most often, the 'evens' player or the 'odds' player, and why they think this was the case.
- Ask the players whether they think the game is fair, and to explain the reasoning behind their answer.

Success criteria

Ask the learners:

- Which of these numbers are odd? 356, 8621, 190, 3172, 509, 2723. How do you know?
- If you add two odd numbers, is the answer odd or even?
- If you subtract an even number from an even number, is the answer odd or even?
- If you add one even number and one odd number, is the answer odd or even?

Ideas for differentiation

Support: In the final Main activity, give these learners 'crib sheets' listing the rules for determining odd and even numbers and the rules for adding and subtracting pairs of numbers.

Extension: In the third Main activity, ask these learners to investigate general statements about subtracting and / or multiplying odd and even numbers.

Number cards

420	281	222	903
174	415	646	807
598	119	350	331
532	683	794	675
3026	1667	9838	2209
8410	4171	9382	2633
1724	2385	5956	7007
6198	3349	8620	4141

Negative numbers 1

Learning objectives

- Use negative numbers in context, e.g. temperature. (4Nn13)
- Explore and solve number problems and puzzles. (4Ps4)
- Use ordered lists and tables to help solve problems systematically. (4Ps5)

Resources

Sets of cards made from photocopiable page 73; A3 paper; photocopiable page 74.

Starter

- Before the lesson, make enough sets of cards from photocopiable page 73 to allow one set between two learners.
- On the board, draw a blank number line with 0 marked in the centre and arrows on either end.
- Give pairs of learners a set of cards made from photocopiable page 73 and a piece of A3 paper. Ask them to copy the number line from the board onto the piece of paper, and then place the cards on the number line in the correct order.

Main activities

- Discuss the placement of the numbers on the number line, demonstrating the correct placement on the board. Explain how to say the numbers with the 'minus' sign in front of them (for example 'negative four'), and introduce the term 'negative numbers' (a negative number is any number less than 0).
- Using the number line, practise counting on and back in 1s, through 0.
- Give the learners pairs of numbers to compare, using the number line to help them. Ask: *Which is greater, 2 or −4? Which is less: −8 or −3?* Emphasise that the numbers to the right of the number line are greater than those on the left.
- Display a copy of photocopiable page 74, discussing the various contexts in which negative numbers are used in the illustrations.
- Set the following puzzle for the learners to solve:

In a quiz, contestants get 1 point for each answer they get right, but lose 1 point for each answer they get wrong. What are all the possible scores a contestant could score by answering 10 questions?

Plenary

- Ask the learners to give all the possible scores that a contestant could get in the quiz. Ask: *How did you make sure you found all the possible scores?*
- Work together as a class to present the calculations in a table, such as the one below. Ask the learners to describe any patterns they can see in the numbers.

Number of correct questions	Number of incorrect questions	Total score
0	10	−10
1	9	−8
2	8	−6
3	7	−4
4	6	−2
5	5	0
6	4	2
7	3	4
8	2	6
9	1	8
10	0	10

Success criteria

Ask the learners:

- What is a negative number?
- Can you describe a situation in which you might need to use negative numbers?
- How did you make sure you found all the possible scores in the quiz?

Ideas for differentiation

Support: In the final Main activity, pair these learners with more confident learners who can provide them with support.

Extension: Ask these learners to investigate all the possible scores when contestants score 2 points for a correct answer and −5 points for an incorrect answer.

Positive and negative numbers

–10	–9	–8	–7
–6	–5	–4	–3
–2	–1	1	2
3	4	5	6
7	8	9	10

Some uses of negative numbers

Look at the illustrations below. How are negative numbers used?

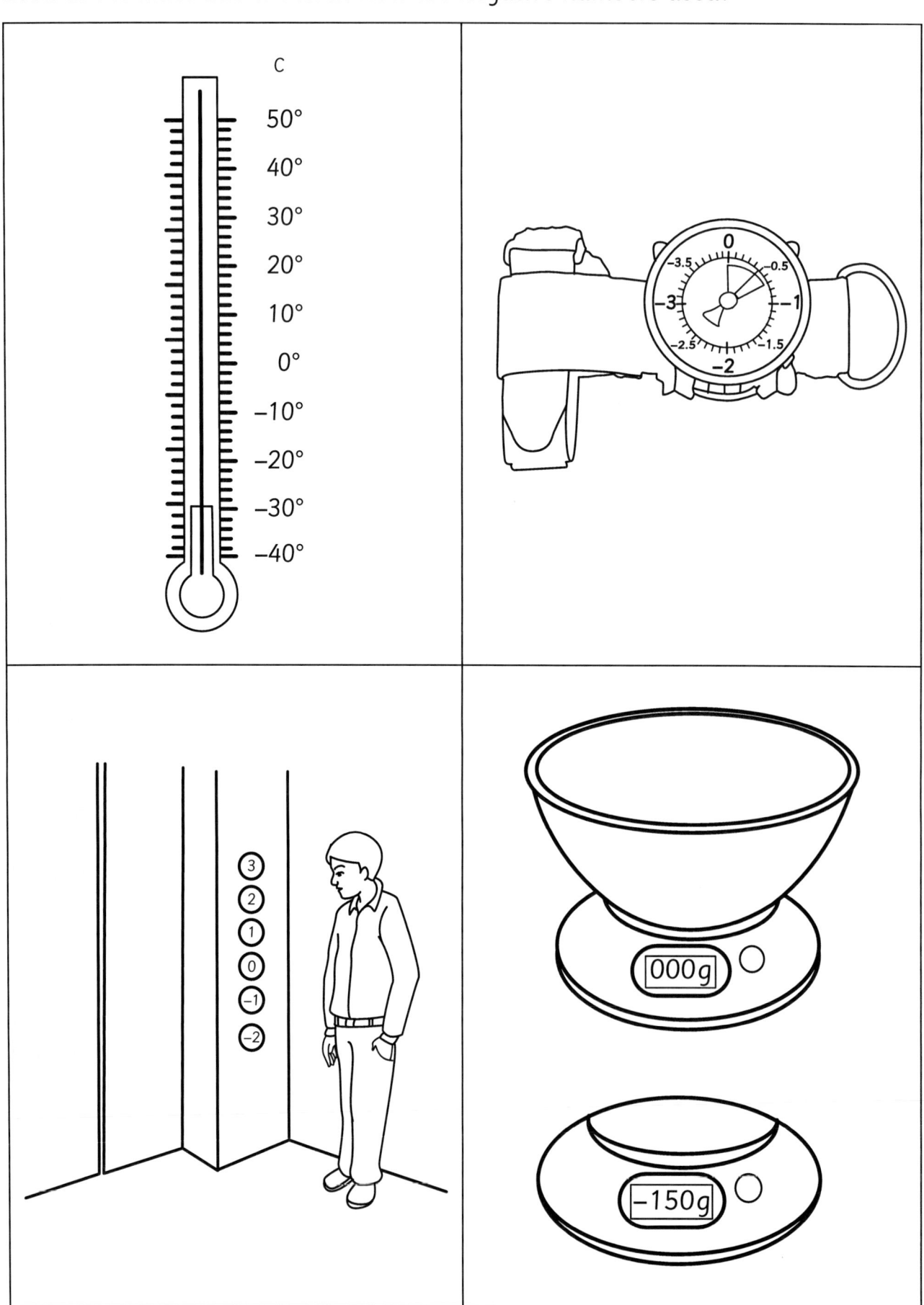

Cambridge Primary: Ready to Go Lessons for Maths Stage 4 © Hodder & Stoughton Ltd 2013

Number sequences 1

Learning objectives

- Recognise and begin to know multiples of 2, 3, 4, 5 and 10 up to the tenth multiple. (4Nc5)
- Describe and continue number sequences, e.g. 7, 4, 1, –2 … identifying the relationship between each number. (4Ps6)
- Recognise and extend number sequences formed by counting in steps of constant size, extending beyond zero when counting back. (4Nn14)

Resources

Counting stick; number lines made from photocopiable page 76; hundred squares including a display copy; counters.

Starter

- Using the counting stick, count on from 0 in steps of 2, for example: *Zero, two, four, six …* Repeat, saying the number of 2s each division represents, for example: *No twos are zero, one two is two, two twos are four, three twos are six …* Point to a random division, for example the seventh, while saying the number of 2s this represents, for example: *Seven twos*. Ask the learners to say the answer (for example 'Fourteen'). Repeat.
- Repeat the activity for counting in 10s, 5s, 3s and 4s.

Main activities

- Give each learner a number line made from photocopiable page 76. On the board, write a sequence of three numbers counting on or back in steps of a constant size whose first six terms are between –20 and 20, for example 3, 1, –1 … Ask the learners to describe the rule (for example 'Count back in twos') and then ask them to write the next three numbers (for example –3, –5, –7). Repeat for several number sequences.

- Display a large hundred square and give each learner a hundred square and a counter. Call out a two-digit starting number. Ask the learners to put their counter on this number. Ask them to count on or back in steps of a given constant size, and write down the first six numbers in the sequence. Repeat for several different number sequences, including jump sizes of 2, 3, 4, 5 and 10.
- Ask the learners to work individually or in pairs to devise their own number sequences, and then give them to a friend to solve and continue.

Plenary

- Ask selected learners to share the number sequences they have devised.
- Ask the rest of the class to describe the rule for each sequence and then write the next three terms in the sequence.

Success criteria

Ask the learners:

- What are six 4s? How did you work out the answer?
- Is 47 a multiple of 5? How do you know?
- Can you describe the rule for this number sequence? 10, 7, 4, 1 …
- What are the next three numbers in the sequence?

Ideas for differentiation

Support: In the final Main activity, pair these learners with a partner who is more confident in Maths.

Extension: In the final Main activity, challenge these learners to devise number sequences that go back beyond –20.

−20 to 20 number lines

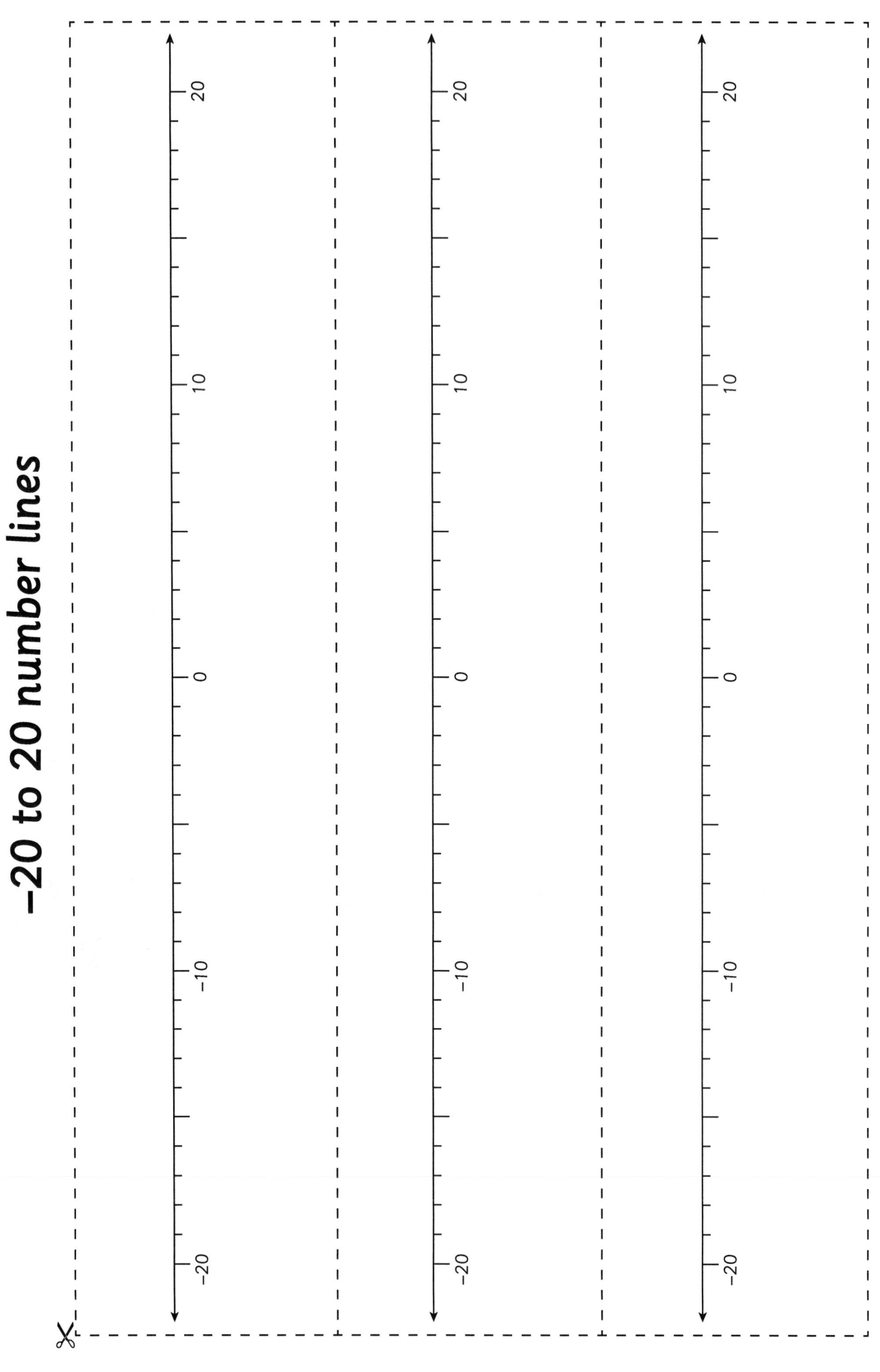

Cambridge Primary: Ready to Go Lessons for Maths Stage 4 © Hodder & Stoughton Ltd 2013

Addition facts

Learning objectives

- Derive quickly pairs of two-digit numbers with a total of 100, e.g. $72 + \square = 100$. (4Nc1)
- Derive quickly pairs of multiples of 50 with a total of 1000, e.g. $850 + \square = 1000$. (4Nc2)

Resources

Counting stick; photocopiable page 78.

Starter

- On a counting stick, count from 0 to 100 in 5s, using each half division to represent 5.
- Point to multiples of 5 on the counting stick in ascending order (5, 10, 15, 20 ...) and ask the learners each time: *How many more to make 100?* (95, 90, 85, 80 ...)
- Point to multiples of 5 on the counting stick in a random order and ask: *What number is this?* (For example sixty-five.) *How many more to make 100?* (For example thirty-five.)

Main activities

- Use knowledge of pairs of multiples of 5 totalling 100 to derive pairs of any two-digit numbers with a total of 100, for example use knowledge that $70 + 30 = 100$ to work out $72 + \square = 100$. Because 72 is 2 more than 70, its complement will be 2 less than 30. So $72 + 28 = 100$. Check by partitioning and recombining (for example $70 + 20 = 90$; $2 + 8 = 10$; $90 + 10 = 100$).

- Use knowledge of pairs of multiples of 5 totalling 100 to derive pairs of multiples of 50 that total 1000, for example $75 + 25 = 100$ so $750 + 250 = 1000$. Check by partitioning and recombining (for example $700 + 200 = 900$; $50 + 50 = 100$; $900 + 100 = 1000$).

- Display photocopiable page 78, work through a few questions and check answers by partitioning and recombining.

- Hand out photocopiable page 78 and ask the learners to complete it, either individually or with a partner.

Plenary

- Discuss the answers to the questions on photocopiable page 78, asking the learners to say which known number fact they used to work out each answer.
- Ask any learners who have devised their own questions to share some of them, and ask the rest of the class to work out the answers. Discuss methods for checking.

Success criteria

Ask the learners:

- What is the missing number in this number sentence? $150 + \square = 1000$
- How could you check your answer?
- How many more do you need to add to 63 to make 100?
- How did you work it out?

Ideas for differentiation

Support: Give these learners a full list of pairs of multiples of 5 that total 100. Ask them to attempt only the first three or four questions in each section on photocopiable page 78.

Extension: Challenge these learners to write their own questions featuring pairs of multiples of 10 that total 1000 (for example $670 + 330 = 1000$).

Name: _____

Complements to 100 and 1000

Fill in the missing number in each number sentence.

Section A

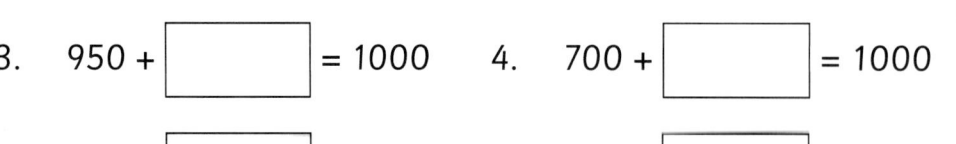

1. 21 + ☐ = 100

2. 84 + ☐ = 100

3. 29 + ☐ = 100

4. 56 + ☐ = 100

5. 33 + ☐ = 100

6. 47 + ☐ = 100

Now write three more questions of your own.

_____

_____

_____

Section B

1. 250 + ☐ = 1000

2. 400 + ☐ = 1000

3. 950 + ☐ = 1000

4. 700 + ☐ = 1000

5. 350 + ☐ = 1000

6 750 + ☐ = 1000

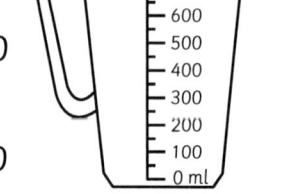

Now write three more questions of your own.

_____

_____

_____

Cambridge Primary: Ready to Go Lessons for Maths Stage 4 © Hodder & Stoughton Ltd 2013

Doubling and halving 1

Learning objectives

- Derive quickly doubles of all whole numbers to 50, doubles of multiples of 10 to 500, doubles of multiples of 100 to 5000, and corresponding halves. (4Nc16)
- Explain methods and reasoning orally and in writing. (4Ps9)

Resources

Beanbag or soft ball; cards made from photocopiable page 80; calculators; timers.

Starter

- Throw a beanbag or soft ball to one learner at a time, asking the learner to answer a doubling question (involving whole numbers to 20) or a related halving question, for example: *What's double eighteen? What's half of twenty-four? What's double fourteen?* The learner must throw the beanbag or ball back to you as they answer the question, and then you throw it to another learner, asking them a different question. Keep the pace as brisk as possible.

Main activities

- On the board write a two-digit number between 20 and 50 that isn't a multiple of 10, for example 48. Ask the learners to double the number and explain their method. Record the answer, for example double 48 = 96.
- On the board, write the first starting number multiplied by 10 (for example 480) and by 100 (for example 4800). Ask the learners to double each number and explain their method. Record the answers, for example double 480 = 960; double 4800 = 9600. Ask: *What pattern do you see?* (For example when the starting number is multiplied by 10, the answer is multiplied by 10.)
- Display a card from photocopiable page 80. Ask the learners to work out the answer, and then check it using the inverse operation. (For example check that half of 54 equals 27 by making sure that double 27 equals 54.) Repeat several times.

- Organise the learners into pairs, giving each pair a set of cards made from photocopiable page 80. The pairs should turn over a card and each learner calculate the answer independently, then check each other's answers using the inverse operation.

Plenary

- On the board, write a two-digit decimal number between 2.1 and 4.9, for example 3.6. Ask the learners to double the number and explain their method.
- Make explicit the link between doubling this number and doubling the number ten times smaller, for example double 36 = 72; double 3.6 = 7.2.

Success criteria

Ask the learners:

- What's double 39? How did you work it out?
- What's half of 560? How did you work it out?
- Which doubling fact can help you work out the answer to double 1200?
- How could you check that half of 92 equals 46?

Ideas for differentiation

Support: In the final Main activity, give these pairs of learners sets of cards with the multiples of 1000 removed, and provide them with calculators for checking answers.

Extension: In the final Main activity, encourage these learners to complete the halving or doubling calculation within a given and challenging time limit.

Doubling and halving cards

double 23	half of 58	double 47	half of 76	double 39
half of 92	double 19	half of 86	double 43	half of 64
double 150	half of 480	double 360	half of 320	double 420
half of 220	double 290	half of 360	double 470	half of 180
double 1700	half of 2600	double 2300	half of 3200	double 3800
half of 4800	double 4100	half of 2400	double 2900	half of 1600

Cambridge Primary: Ready to Go Lessons for Maths Stage 4 © Hodder & Stoughton Ltd 2013

Addition and subtraction strategies 1

Learning objectives

- Add any pair of two-digit numbers, choosing an appropriate strategy. (4Nc9)
- Subtract any pair of two-digit numbers, choosing an appropriate strategy. (4Nc10)
- Choose appropriate mental or written strategies to carry out calculations involving addition and subtraction. (4Pt1)

Resources

A set of large cards made from photocopiable page 82; ten-sided dice; modified six-sided dice with three faces marked with a + and three marked with a −.

Starter

- Write the following number sentences on the board, explaining that each box stands for a missing digit. Ask the learners to rewrite the number sentences, filling in the missing digits. Challenge them to find as many different answers as they can in the time available. Discuss their answers.

$$\Box + \Box = 9 \qquad \Box - \Box = 2$$
$$\Box 0 + \Box 0 = 120 \qquad \Box 0 - \Box 0 = 30$$

Main activities

- Display a large card made from photocopiable page 82. Ask the learners to do the calculation and record their answer. Discuss methods used. Repeat for other cards.
- Discuss the fact that different methods may suit different calculations. Methods may include:
  - partitioning into tens and units and then recombining (for example Card 4)
  - starting with the larger number and then counting on, either with or without a number line (for example Card 1)
  - looking for doubles and near doubles (for example Cards 2 and 5)
  - rounding up or down to a multiple of 10 and then adjusting (for example Cards 3, 7 and 8).

- Organise the learners into groups of four to six. One member of the group should roll a ten-sided dice four times to generate two two-digit numbers, and then roll the modified six-sided dice once to generate an addition or subtraction. Each group member must calculate the answer using any method they like. When everyone has finished calculating, they should compare answers, discuss methods and try to agree on the correct answer. Everyone who has the correct answer scores a point.

Plenary

- Describe a method of calculation discussed during the lesson (for example rounding up or down to a multiple of 10 and then adjusting). Ask the learners in pairs to make up a calculation that would be suitable to solve using this method.

Success criteria

Ask the learners:

- What is 35 + 36?
- How did you work out the answer?
- What is 47 − 19?
- How did you work out the answer?

Ideas for differentiation

Support: During the final Main activity, group these learners together and give them a six-sided dice for generating two-digit numbers instead of a ten-sided dice.

Extension: During the final Main activity, group these learners together and introduce a time limit for each round.

Addition and subtraction cards

Card 1

78 + 16

Card 2

25 + 26

Card 3

53 – 29

Card 4

86 – 42

Card 5

44 + 43

Card 6

72 – 25

Card 7

93 – 58

Card 8

39 + 54

Cambridge Primary: Ready to Go Lessons for Maths Stage 4 © Hodder & Stoughton Ltd 2013

Addition and subtraction strategies 2

Learning objectives

- Add three two-digit multiples of 10, e.g. 40 + 70 + 50. (4Nc7)
- Add and subtract near multiples of 10 or 100 to or from three-digit numbers, e.g. 367 − 198 or 278 + 49. (4Nc8)
- Find a difference between near multiples of 100, e.g. 304 − 296. (4Nc11)
- Subtract a small number crossing 100, e.g. 304 − 8. (4Nc12)

Resources

Two sets of large cards of multiples of 10 up to 90; photocopiable page 84.

Starter

- Shuffle together the two sets of large cards of multiples of 10 up to 90. Draw three cards and display them (for example 40, 40, 60). Ask the learners to add the three numbers together and write down the total.
- Discuss strategies used. These might include:
  - looking for pairs that make 100 (for example 40 + 60)
  - using doubles knowledge (for example double 40 is 80)
  - using knowledge of single-digit addition facts and multiplying by 10 (for example 4 + 4 + 6 = 14 so 40 + 40 + 60 = 140).
- Repeat for other sets of three cards.

Main activities

- Write a calculation on the board that involves subtracting a single-digit number from a three-digit number that crosses the hundreds boundary (for example 503 − 7). Ask the learners to find the answer and explain their method. Model strategies, including counting back on a number line.
- Give the learners other similar calculations, for example 304 − 6; 705 − 8; 201 − 7; 402 − 9. Ask: *How could you check your answer?* Model adding the answer to the smaller number in the original calculation.

- Write a calculation on the board that involves a small difference between near multiples of 100, for example 303 − 298. Ask: *Would starting at 303 and counting back 298 be a good strategy?* (No.) *Why not?* (Because it would take a long time and you could easily lose count.) Ask the learners to suggest an alternative strategy (for example locating 303 and 298 on a number line and finding the difference). Give the learners other similar calculations, for example 507 − 492; 701 − 693; 203 − 197; 406 − 399.
- Give out photocopiable page 84, asking the learners to work through the questions either individually or with a partner.

Plenary

- Discuss answers to the questions on photocopiable page 84.
- For some of the questions, ask the learners to describe the method they used. For other questions, ask the learners to say what calculation they could do in order to check the answer.

Success criteria

Ask the learners:

- What's 30 + 70 + 50? How did you work out the answer?
- What's 304 − 8? How did you work out the answer?
- What's 304 − 296? How did you work out the answer?
- What's 278 + 49? How did you work out the answer?

Ideas for differentiation

Support: Group these learners together and work through the first three or four questions on photocopiable page 84 with them.

Extension: Ask these learners to write a number story to go with each calculation on photocopiable page 84.

Name: _____

Addition and subtraction strategies 2

Answer the questions below. Show your workings in the box under each sum.

1.   304 – 5 = _____

2.   403 – 391 = _____

3.   245 + 49 = _____

4.   573 – 298 = _____

5.   806 – 9 = _____

6.   602 – 596 = _____

7.   773 + 61 = _____

8.   236 + 53 = _____

Cambridge Primary: Ready to Go Lessons for Maths Stage 4 © Hodder & Stoughton Ltd 2013

More addition

Learning objectives

- Make up a number story for a calculation. (4Ps1)
- Add pairs of three-digit numbers. (4Nc17)
- Check the results of adding numbers by adding them in a different order or by subtracting one number from the other. (4Pt3)
- Estimate and approximate when calculating and check working. (4Pt8)

Resources

Large cards made from photocopiable page 86; cards made from photocopiable page 12, including a large set for display.

Starter

- Display one of the large cards made from photocopiable page 86. Ask the learners to work individually or with a partner to make up a number story to go with the calculation (for example for the calculation $96 \div 12 = 8$: 'There are 96 tennis balls in the games cupboard. Each pack of tennis balls contains 12 balls, and there are 8 packs altogether.')
- Share number stories. Repeat the activity for each of the remaining cards.

Main activities

- Shuffle a set of number cards made from photocopiable page 12. Draw six cards to generate two three-digit numbers.
- Ask the learners to estimate the total (for example an estimate for $196 + 462$ might be $200 + 460 = 660$).
- Ask the learners to calculate the total and explain their method. Discuss methods used. Methods may include:
  - partitioning and recombining
  - counting on from the larger number (with or without a number line)
  - rounding up or down to the nearest 10 or 100 and adjusting
  - looking for doubles and near doubles.
- Ask: *How can you check your answer?* (For a rough check compare your answer to the estimate; for a closer check add the numbers in a different order, or subtract one number from the other.)

- Organise the learners into groups of four to six. Give each group a set of cards made from photocopiable page 12. The learners must use the cards to generate pairs of three-digit numbers, which they add together, then check each other's answers.

Plenary

- On the board, write some three-digit additions with answers, some of which are incorrect, for example:

1.

		5	7	7
+		4	3	8
=	1	0	1	5

2.

		8	9	9
+		2	1	7
=	1	0	1	6

3.

	1	9	4
+	7	2	0
=	9	1	4

4.

	6	2	1
+	1	8	9
=	8	1	0

5.

	3	1	0
+	3	8	3
=	6	9	2

- Challenge the learners to identify which answers are correct and which are incorrect. (In the examples above, 2 and 5 are incorrect.) Ask the learners to explain how they checked the answers.

Success criteria

Ask the learners:

- Can you make up a number story to go with this calculation: $180 \div 9 = 20$?
- Estimate the answer to $375 + 522$. How did you reach your estimate?
- What is $375 + 522$? How did you work it out?
- How could you check your answer?

Ideas for differentiation

Support: In the final Main activity, organise these learners to work in pairs within their group.

Extension: In the final Main activity, group these learners together. When they have had some practice adding pairs of three-digit numbers, challenge them to add three three-digit numbers.

Number story calculation cards

$$27 + 56 + 38 = 121$$

$$98 - 54 + 19 = 63$$

$$35 \times 7 = 245$$

$$96 \div 12 = 8$$

 Cambridge Primary: Ready to Go Lessons for Maths Stage 4 © Hodder & Stoughton Ltd 2013

More subtraction

Learning objectives

- Subtract a two-digit number from a three-digit number. (4Nc18)
- Subtract pairs of three-digit numbers. (4Nc19)
- Choose strategies to find answers to addition or subtraction problems; explain and show working. (4Ps3)

Resources

Place value cards (hundreds, tens and units) including one large set for display; photocopiable page 88.

Starter

- Organise the learners into pairs, giving each pair a set of place value cards. Write a three-digit number on the board, for example 581, and ask the learners to make it using their place value cards. Ask: *Which cards did you use?* (For example 500, 80 and 1.) Repeat for other three-digit numbers, including some containing a 0.
- Display a three-digit number on the large place value cards, for example 347, and ask: *Which cards did I use?* (For example 300, 40 and 7.) Confirm the answer by pulling the place value cards apart. Repeat for other three-digit numbers, including some containing a 0.

Main activities

- On the board, write a calculation in which a two- or three-digit number is subtracted from a three-digit number, with no exchanging of tens or units, for example 467 − 32. Ask the learners to work out the answer, recording their working, and explaining their method. Methods may include partitioning and recombining, counting back (with or without a number line) or rounding to the nearest 10 or 100 and adjusting. Repeat for similar subtractions.

- Ask the learners to subtract a two- or three-digit number from a three-digit number where exchanging tens or ones is required, for example 224 − 68. Any learners who use partitioning and recombining may flounder on the calculations 20 − 60 and 4 − 8. Demonstrate exchanging one hundred for ten tens and one ten for ten units in order to facilitate the subtraction. Give the learners other similar subtractions.

- Hand out photocopiable page 88. Explain how the subtraction wall works. Each number in the wall is the difference between the two numbers above it. Work through one calculation together.

Plenary

- Ask the learners to give the number in the bottom brick in the subtraction wall.
- Ask the learners to give the rest of the answers for the subtraction wall.

Success criteria

Ask the learners:

- What's 298 − 53? How did you work it out?
- What's 437 − 215? How did you work it out?
- What's 334 − 76? How did you work it out?
- There are 500 tickets available for a concert. 236 tickets have been sold. How many tickets are left? How did you work out the answer?

Ideas for differentiation

Support: Group these learners together and work through a few more subtractions in the subtraction wall with them.

Extension: Ask these learners to extend the subtraction wall upwards, suggesting a row of six numbers that could go above the top row.

Subtraction wall

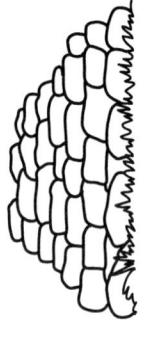

In this subtraction wall, each number is the difference between the two numbers above it.

Fill in the missing numbers in the wall.

465	319	244	582	803
146	75	338	221	
71	263	117		
192	146			
46				

Cambridge Primary: Ready to Go Lessons for Maths Stage 4 © Hodder & Stoughton Ltd 2013

Multiplication and division facts

Learning objectives

- Know multiplication for 2×, 3×, 4×, 5×, 6×, 9× and 10× tables and derive division facts. (4Nc4)
- Multiply any pair of single-digit numbers together. (4Nc13)
- Check multiplication using a different technique, e.g. check 6 × 8 = 48 by doing 6 × 4 and doubling. (4Pt5)

Resources

Cards made from photocopiable pages 90 and 91; timer; ten-sided dice; calculators; six-sided dice.

Starter

- Shuffle the cards made from photocopiable pages 90 and 91, keep the card marked 'START' and hand out the rest to the learners. Some learners may need more than one card. Call out the multiplication on the 'START' card, and ask the learner with the answer to read it out and then read out the multiplication on the bottom of their card. Have the learners give the answers and questions in the same way until you reach the 'END' card.
- Redistribute the cards and repeat the activity, challenging the learners to complete it in a given time limit (for example three minutes).

Main activities

- Generate two single-digit numbers by throwing two ten-sided dice. Ask the learners to multiply the numbers together.
- Take learners' answers, asking them to explain their methods. Methods may include 'just knowing' the answer, drawing sets or an array, or using times table knowledge, for example working out 7 × 6 by doubling 7 × 3, or by adding 4 × 6 and 3 × 6. Methods using times tables knowledge are likely to be the most varied. Model checking answers by using a different method of multiplication. Repeat for several other pairs of single-digit numbers.

- Organise the learners into pairs, giving each pair two ten-sided dice. Ask pairs to use the dice to generate two single-digit numbers, multiply them together (recording any workings) and write down the answer. The learners should check each other's answers using a different technique.

Plenary

- Ask the learners in pairs to calculate the answer to 12 × 6.
- Take answers, asking the learners to explain how they did the calculation. Challenge them to find as many different methods as possible.

Success criteria

Ask the learners:

- What's 3 × 9?
- What's 40 ÷ 5? How did you work out the answer?
- What's 7 × 8? How did you work out the answer?
- How could you check your answer to 7 × 8?

Ideas for differentiation

Support: In the final Main activity, ask these learners to generate two single-digit numbers by throwing six-sided dice instead of ten-sided dice.

Extension: In the last few minutes of the final Main activity, challenge these learners to multiply three single-digit numbers together.

Times table follow-me cards 1

START	20	7	32
What is 10 × 2?	What is 21 ÷ 3?	What is 8 × 4?	What is 45 ÷ 5?
9	30	2	90
What is 5 × 6?	What is 18 ÷ 9?	What is 9 × 10?	What is 24 ÷ 4?
6	70	8	25
What is 7 × 10?	What is 16 ÷ 2?	What is 5 × 5?	What is 12 ÷ 4?
3	27	4	35
What is 3 × 9?	What is 24 ÷ 6?	What is 7 × 5?	What is 9 ÷ 9?
1	12	10	14
What is 2 × 6?	What is 30 ÷ 3?	What is 7 × 2?	What is 20 ÷ 4?

Cambridge Primary: Ready to Go Lessons for Maths Stage 4 © Hodder & Stoughton Ltd 2013

Times table follow-me cards 2

5	**100**	**36**	**40**
What is 10 × 10?	What is 4 × 9?	What is 4 × 10?	What is 8 × 3?
24	**45**	**22**	**18**
What is 5 × 9?	What is 11 × 2?	What is 3 × 6?	What is 3 × 5?
15	**16**	**21**	**28**
What is 8 × 2?	What is 7 × 3?	What is 7 × 4?	What is 10 ÷ 5?
50	**42**	**81**	**33**
What is 7 × 6?	What is 9 × 9?	What is 3 × 11?	What is 0 × 4?
0	**60**	**48**	**54**
What is 6 × 10?	What is 8 × 6?	What is 6 × 9?	END

Multiplication strategies 5

Learning objectives

- Use knowledge of commutativity to find the easier way to multiply. (4Nc14)
- Multiply a two-digit number by a single-digit number. (4Nc22)
- Explain reasons for a choice of strategy when multiplying or dividing. (4Ps2)

Resources

Multiplication grid from photocopiable page 93; number cards made from photocopiable page 12.

Starter

- Organise the learners into pairs and give each pair a multiplication grid from photocopiable page 93. Ask the learners to look carefully at the numbers in the multiplication grid, and talk with their partner about any patterns they can see.
- Ask the learners to describe the patterns they have found. Patterns of digits can be seen within the number sequences in individual rows and columns; also, each row has a corresponding identical column. Some learners may have noticed the symmetry of the numbers in the entire grid, with the diagonal that extends from top left to bottom right acting as a 'mirror line'.

Main activities

- On the board write a multiplication involving one two-digit and one single-digit number, for example 25 × 3. Ask: *What's the answer? How did you work it out? Why did you choose that way of doing it?* Discuss methods used, which may include repeated addition, or partitioning and recombining (either with or without a multiplication grid like those on photocopiable page 93). Discuss the fact that the order of multiplying numbers makes no difference to the answer (for example 25 × 3 = 3 × 25). Use this knowledge of commutativity to find the easier way to multiply, for example 3 lots of 25 (25 × 3) may be easier than 25 lots of 3 (3 × 25).

- Give the learners another multiplication, for example 37 × 5. Ask them to give an estimate first, and describe their estimation method.
- Give them another multiplication, for example 42 × 6. Ask them how to check their answers (for example by using a different method of multiplication).
- Organise the learners into pairs, giving each pair a set of number cards made from photocopiable page 12. Ask pairs to use the cards to generate one two-digit number and one single-digit number, and multiply them, recording their workings. The learners should check each other's answers using a different method.

Plenary

- On the board, write a multiplication involving one three-digit number and one single-digit number (for example 265 × 9). Ask the learners to work in pairs to do the calculation, and then explain their method.

Success criteria

Ask the learners:

- If you know 7 × 6 = 42, what other times table fact must you know?
- What is 45 × 8?
- How did you work out the answer?
- Why did you choose that way of doing the calculation?

Ideas for differentiation

Support: In the final Main activity, give groups of these learners a calculator for checking answers.

Extension: In the final Main activity, give groups of these learners a set of number cards made from two lots of the numbers 5 to 9.

Multiplication grid

×	0	1	2	3	4	5	6	7	8	9	10
0	0	0	0	0	0	0	0	0	0	0	0
1	0	1	2	3	4	5	6	7	8	9	10
2	0	2	4	6	8	10	12	14	16	18	20
3	0	3	6	9	12	15	18	21	24	27	30
4	0	4	8	12	16	20	24	28	32	36	40
5	0	5	10	15	20	25	30	35	40	45	50
6	0	6	12	18	24	30	36	42	48	54	60
7	0	7	14	21	28	35	42	49	56	63	70
8	0	8	16	24	32	40	48	56	64	72	80
9	0	9	18	27	36	45	54	63	72	81	90
10	0	10	20	30	40	50	60	70	80	90	100

Division strategies 2

Learning objectives

- Divide two-digit numbers by single-digit numbers. (4Nc23)
- Decide whether to round up or down after division to give an answer to a problem. (4Nc24)
- Check the result of a division using multiplication, e.g. multiply 4 by 12 to check 48 ÷ 4. (4Pt6)

Resources

Beanbag or soft ball; photocopiable page 95; multiplication grid from photocopiable page 93.

Starter

- Throw a beanbag or soft ball to one learner at a time, asking the learner a times tables division question, for example: *What's 24 ÷ 4?* Give the average learners questions from the 3 and 4 times tables, the lower-ability learners questions from the 2, 5 and 10 times tables, and the more-able learners questions from the 6 and 9 times tables.
- The learner should throw the beanbag or ball back to you as they answer the question, and then you throw it to another learner, asking them a different question. Keep the pace as brisk as possible.

Main activities

- Write 76 ÷ 3 on the board. Ask the learners how to do the calculation (for example partition 76 into 60 + 16, divide each number separately and then recombine). Perform the calculation, and then ask: *How could you check the answer?* (For example check that 76 ÷ 3 = 25 r1 by doing 25 × 3 and then adding 1 to see if the answer is 76.)
- Express 76 ÷ 3 in the form of two word problems requiring a whole number answer. One problem should require rounding up and one rounding down, for example:
  - *You have 76 balloons to put into packs of three. How many full packs can you make?*

- *You need 76 cupcakes for a party. Cupcakes are sold in boxes of three. How many boxes of cupcakes do you need to buy?* Say: *We already know the answer (25 r1), but both problems need a whole number answer. For each problem, ask: Should you round the answer up (to 26) or down (to 25)? Why?*
- Hand out photocopiable page 95, and work through the first question together. Ask the learners to work through the rest of the questions individually or with a partner.

Plenary

- Go through the answers to photocopiable page 95. Ask the learners to explain their division strategies, and describe how to check the answer using multiplication and addition.
- For each word problem, ask the learners to say whether they rounded the answer up or down, and explain why.

Success criteria

Ask the learners:

- What's 63 ÷ 5?
- How did you work out the answer?
- How could you check that the answer is correct?
- How many 5-litre buckets can be filled from a water barrel containing 63 litres? Will you need to round the answer up or down? Why?

Ideas for differentiation

Support: Pair these learners with a more confident partner, and give them a multiplication square to support division.

Extension: Challenge these learners to write their own pairs of word problems featuring the same division calculation, one of which requires rounding up and one of which requires rounding down.

Name: _____

Round up or round down?

Do the calculation in part a) of each question.
Record your working and write your answer on this page.

The word problems in part b) and c) match the calculation in part a),
but they need a whole number answer. Write the whole number answers.

Question 1

a) 88 ÷ 7 = _____ r_____

b) You want to make a scarf 88 cm long by joining fabric squares end to end.

 Each square is 7 cm long. How many squares do you need? _____

c) How many whole weeks are there in 88 days? _____

Question 2

a) 68 ÷ 6 = _____ r_____

b) A chef needs 68 eggs. Eggs come in boxes of six.

 How many boxes of eggs does the chef need to buy? _____

c) Six friends share 68 $1 coins between them.

 How many $1 coins does each friend get? _____

Question 3

a) 73 ÷ 4 = _____ r_____

b) A gardener has 73 plants to plant in four equal rows.

 How many plants can the gardener plant in each row? _____

c) A team of 73 athletes is travelling to a sporting event by car.

 Each car can take four passengers.

 What is the minimum number of cars needed to transport the athletes?

 _____

Unit assessment

Questions to ask

- What place value cards would you need to make the number 3056?
- What is 372 rounded to the nearest hundred?
- What are the next three numbers in this number sequence? 11, 8, 5, 2 …
- What is the missing number in this number sentence: $350 + \square = 1000$?
- What rules do you know about adding and subtracting odd and even numbers?
- What is 429 multiplied by 10? Divided by 10?

Summative assessment activities

Observe the learners while they take part in these activities. You will quickly be able to identify those who appear to be confident and those who may need additional support.

What's the target?

This game assesses the learners' ability to add and subtract numbers with up to three digits.

You will need:

0 to 9 digit cards; timers.

What to do

- Organise the learners into pairs. Give each pair a set of 0 to 9 digit cards and ask them to use the cards to generate a three-digit number. They should set the timer for an agreed time, and each learner must write as many additions and subtractions as possible, each made up of two numbers with at least two digits each, that make the target number.

- When the time is up, the learners should check each other's calculations, and score 1 point for every correct addition or subtraction.

- Repeat for other randomly generated three-digit target numbers. To make the game more challenging you can ban multiples of 10 from the calculations.

Make a set of 'follow-me' cards

This activity assesses the learners' times table knowledge.

You will need:

Follow-me cards made from photocopiable pages 90 and 91; rulers; scissors; card.

What to do

- Remind the learners of how a set of follow-me cards works by working through the set of cards made from photocopiable pages 90 and 91.

- Organise the learners into pairs, giving each pair rulers, scissors and card. Tell learners that you want them to write their own set of between 12 and 20 follow-me cards. Allocate each pair two times tables, appropriate to their ability. Tell them they should write a mix of multiplication and division questions, reminding them that for the cards to work, each answer must only appear once (for example you can't use both 5×2 and $20 \div 2$, because both answers are 10).

- At the end of the session, give the learners the opportunity to test out sets of finished cards.

Written assessment

- Distribute photocopiable page 97. Ask the learners to read the questions and write the answers. They should work independently.

Name: _____

Find the answers

1. Order these temperatures from lowest to highest:

   12°C 0°C 23°C −9°C 36°C −4°C 7°C −1°C

   _____

2. Round each number to the nearest 10 and then to the nearest 100.

   a) 293 _____   b) 4037 _____   c) 526 _____

   _____   _____   _____

3. Multiply each number by 10 and then by 100.

   a) 245 _____   b) 671 _____   c) 304 _____

   _____   _____   _____

4. What number will come out of the function machine when each of these numbers is put in?

   a) 590 _____

   b) 340 _____

   c) 824 _____

   d) 739 _____

   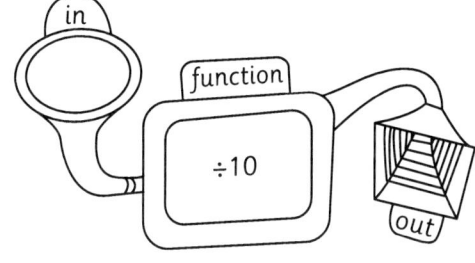

5. Find the doubles and halves.

   a) double 47 _____   b) half of 78 _____

   c) double 360 _____   d) half of 420 _____

   e) double 2500 _____   f) half of 3800 _____

6. 64 x 4 = _____

7. Bouncy balls cost $6 each. You have $70. How many bouncy balls can you buy?

   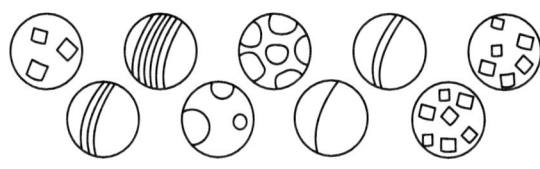

   _____

Unit 2B: Geometry and problem solving

2D shapes 1

Learning objectives

● Identify, describe, visualise, draw and make a wider range of 2D and 3D shapes including a range of quadrilaterals, the heptagon and tetrahedron; use pinboards to create a range of polygons. Use spotty paper to record results. (4Gs1

Resources

Pinboards; elastic bands; dotted paper (square grid); cards made from photocopiable page 99.

Starter

- Introduce the term 'polygon' to describe a closed 2D shape with three or more sides, all of which are straight.

- Organise the learners into pairs, giving each pair a pinboard and an elastic band. Call out the name of a polygon, for example triangle, rectangle, square, hexagon, octagon, pentagon, and ask the learners to make an example of that type of polygon on their pinboard. Introduce the term 'heptagon' for a polygon with seven sides.

Main activities

- Demonstrate using the dotted paper how to record a polygon that has been made on the pinboard. Ask each pair to make and record as many different polygons as they can in a given time, for example 10 minutes.

- Ask the learners to cut out their dotted paper polygons and sort them into groups using their own criteria (for example the number of sides, the number of equal sides, the number of lines of symmetry, or whether they contain a right angle).

- Ask them to describe how they have sorted the shapes, and say how many of each type of shape they have found.

- Ask them to identify the triangles and name as many of them as they can, revising the terms 'equilateral', 'isosceles', 'scalene' and 'right-angled'.

- Revise the term 'quadrilateral' for four-sided polygons. Ask the learners to identify the quadrilaterals and name as many of them as they can, revising the terms 'parallelogram', 'rhombus' and 'kite'.

- Give each pair of learners a set of cards made from photocopiable page 99. Ask them to match each picture with a name and a list of properties. Ask finishers to create their own sets of polygon matching cards for others to match.

Plenary

- Confirm which cards match each other from photocopiable page 99.

- Ask the learners to sketch on dotted paper as many polygons as they can with a given property in a short time limit, for example two minutes. Ask: *How many polygons can you sketch that have no right angles / two equal sides / just one line of symmetry?*

Success criteria

Ask the learners:

● What is a polygon?
● Can you make an isosceles triangle on this pinboard?
● Can you draw the isosceles triangle you made on the pinboard on dotted paper?
● (Pointing to a polygon picture card from photocopiable page 99:) Can you name and describe this shape?

Ideas for differentiation

Support: Group these learners together for the final Main activity and work with them, helping them to read, interpret and match the cards.

Extension: In the final Main activity, ask these learners to make a fourth card for each shape and describe its symmetrical properties.

Polygon matching cards

	isosceles triangle	a three-sided polygon with one pair of equal sides
	equilateral triangle	a three-sided polygon whose sides are all the same length and whose angles are all equal
	scalene triangle	a three-sided polygon whose sides are all different lengths
	parallelogram	a polygon with two pairs of equal sides; equal sides are parallel to each other
	rhombus	a polygon with four equal sides whose angles are not right angles
	kite	a polygon with two pairs of equal sides; equal sides are next to each other
	heptagon	a polygon with seven sides

2D shapes 2

Learning objectives

- Identify, describe, visualise, draw and make a wider range of 2D and 3D shapes including a range of quadrilaterals, the heptagon and tetrahedron; use pinboards to create a range of polygons. Use spotty paper to record results. (4Gs1)
- Classify polygons (including a range of quadrilaterals) using criteria such as the number of right angles, whether or not they are regular and their symmetrical properties. (4Gs2)
- Recognise the relationships between 2D shapes and identify the differences and similarities between 3D shapes. (4Pt7)

Resources

2D shapes, e.g. circles, semi-circles and various triangles, quadrilaterals, pentagons, hexagons, heptagons and octagons; A3 paper; sticky notes; photocopiable page 101.

Starter

- Display about ten 2D shapes, giving each shape a unique identifying number. Ask the learners to identify various subsets within the shapes, for example: *Which shapes are polygons? Which are quadrilaterals? Which shapes have two or more equal sides? Which shapes have at least one right angle?*

Main activities

- Leaving the 2D shapes on display, draw a two-criteria Carroll diagram on the board, in which the columns are labelled 'polygon' and 'not a polygon' and the rows are labelled 'at least one pair of equal sides' and 'no equal sides'.
- Ask the learners to write the number of each shape in the correct region of the diagram.
- Rub out the identifying numbers and the labels on the rows and columns. Write a new set of identifying numbers in each region of the diagram, meeting a set of criteria that you keep secret from the learners, for example 'one or more right angles' / 'no right angles' and 'four or fewer sides' / 'more than four sides'.

- Challenge the learners to work out what the secret labels are, discussing their ideas with a partner. Discuss answers.
- Organise the learners into pairs, giving each pair A3 paper, sticky notes and a selection of polygons. Ask them to sort the shapes according to their own criteria, which they should write as labels on sticky notes, then draw a Carroll diagram and record the shapes' placement by drawing around them and then remove the sticky notes. They should give their Carroll diagram to another pair to try to work out the missing labels.

Plenary

- Ask the learners to sketch a quadrilateral with equal sides that are parallel to each other (for example a parallelogram, rectangle or square).
- Explain that a rectangle and a square are special types of parallelogram. Ask the learners which properties of each set it apart from other parallelograms.

Success criteria

Ask the learners:

- Which of these shapes are polygons?
- Which of the polygons are quadrilaterals?
- Which of the quadrilaterals have right angles?
- Which region of the Carroll diagram will you put this shape in? Why?

Ideas for differentiation

Support: In the final Main activity, provide these learners with photocopiable page 101, which gives ideas for sorting shapes into a Carroll diagram.

Extension: In the final Main activity, group these learners together and challenge them to choose and draw their own 2D shapes.

Pairs of labels

Choose from the following labels to sort the shapes on your Carroll diagram.

One or more right angles / No right angles

Four right angles / Fewer than four right angles

At least one pair of equal angles / No equal angles

At least one pair of equal sides / No equal sides

Four or more equal sides / Fewer than four equal sides

All sides equal / Not all sides equal

All angles equal / Not all angles equal

At least one pair of parallel sides / No parallel sides

Five or more sides / Fewer than five sides

At least one curved side / No curved sides

At least one line of symmetry / No lines of symmetry

3D shapes 1

Learning objectives

● Identify, describe, visualise, draw and make a wider range of 2D and 3D shapes including a range of quadrilaterals, the heptagon and tetrahedron; use pinboards to create a range of polygons. Use spotty paper to record results. (4Gs1)

● Recognise the relationships between 2D shapes and identify the differences and similarities between 3D shapes. (4Pt7)

Resources

3D shapes, e.g. sphere, hemisphere, cone, cylinder, cube, square-based cuboid, triangular prism, hexagonal prism, octagonal prism, triangular-based pyramids including a tetrahedron, square-based pyramid, pentagonal-based pyramid, hexagonal-based pyramid; cards made from photocopiable pages 103 and 104.

Starter

• Before the lesson, give each learner a solid shape with a unique identifying number.

• Introduce the term 'polyhedron' (a solid shape whose surfaces are all flat and whose edges are all straight).

• Display the solid shapes, asking the learners to sort them into two groups: 'polyhedra' and 'not polyhedra'.

• Ask the learners to write the name of each shape. Discuss answers, introducing the term 'tetrahedron' for a triangular-based pyramid in which all the faces are identical equilateral triangles.

Main activities

• Revise the terms 'face' (a flat surface), 'edge' and 'vertex' (the point where three or more faces meet).

• Describe one of the shapes on display in terms of the number of its edges and vertices and / or the number and shape of its faces. Ask the learners to identify the shape. Repeat for other shapes.

• Hold up two shapes, asking the learners to describe how they are the same and how they are different. Repeat for other pairs of shapes.

• Display three shapes that have a property in common, but do not say what this property is. Ask the learners to describe how the shapes are similar.

• Organise the learners into pairs and provide each pair with a set of cards made from photocopiable pages 103 and 104. Tell the learners to take it in turns in their pairs to choose three cards that show shapes with a common property, and ask their partner to say what the shapes have in common.

Plenary

• Display four shapes, three of which share a property, for example a square-based pyramid, a tetrahedron, a square-based cuboid and a cube.

• Ask the learners to say which shape is the odd one out and why (for example the tetrahedron, because it doesn't have any square faces).

Success criteria

Ask the learners:

● What is a polyhedron?

● What is this shape called? What can you tell me about it?

● What is the same about these shapes? What is different about them?

● Which shape is a tetrahedron? What is special about a tetrahedron?

Ideas for differentiation

Support: Group these learners together for the final Main activity and give them more practice guessing common properties of groups you have made, before making their own groupings.

Extension: Challenge these learners to select four shapes, three of which share a property. Their partner has to say which shape is the odd one out and why.

3D shape cards 1

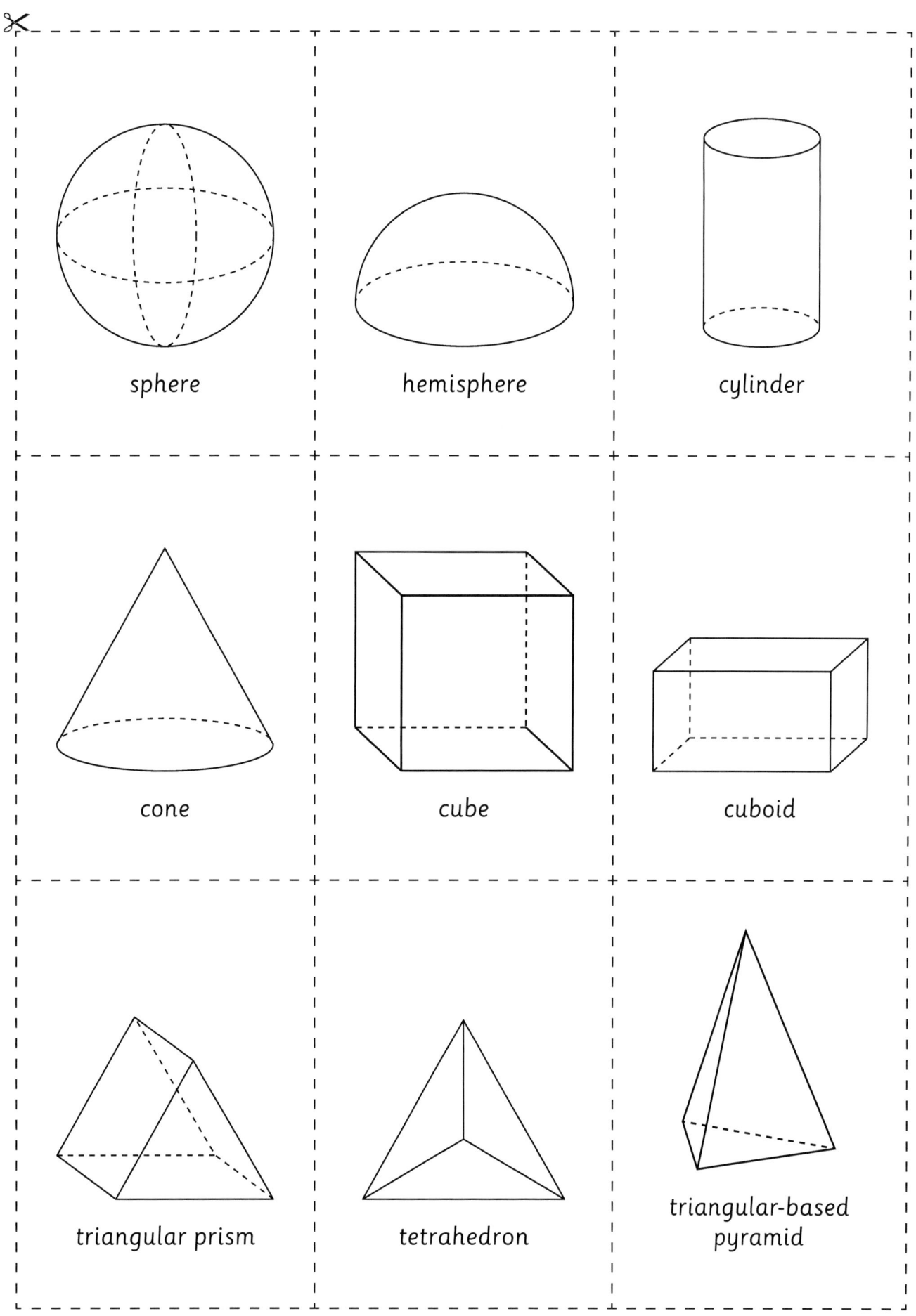

sphere

hemisphere

cylinder

cone

cube

cuboid

triangular prism

tetrahedron

triangular-based pyramid

3D shape cards 2

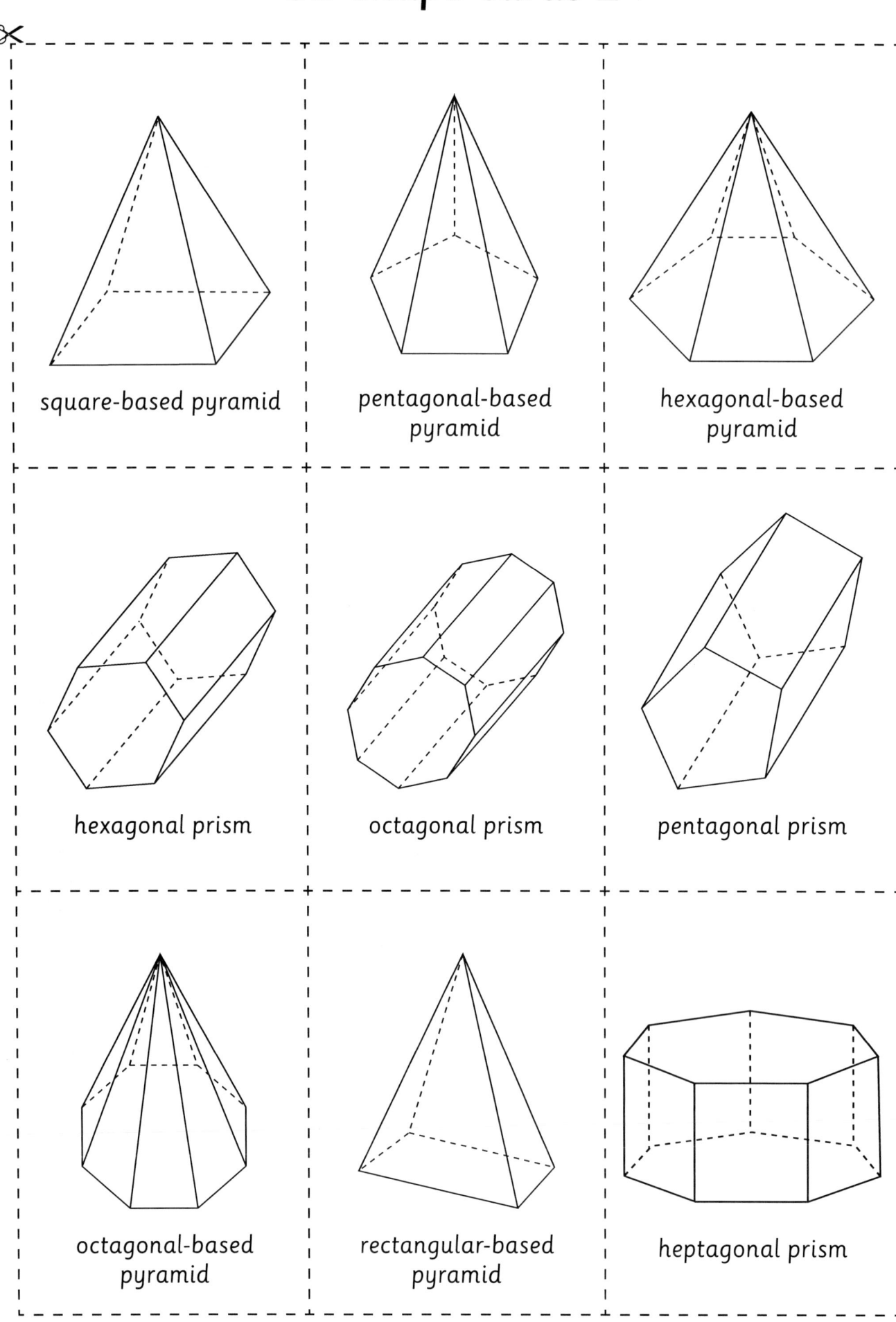

square-based pyramid

pentagonal-based pyramid

hexagonal-based pyramid

hexagonal prism

octagonal prism

pentagonal prism

octagonal-based pyramid

rectangular-based pyramid

heptagonal prism

 Cambridge Primary: Ready to Go Lessons for Maths Stage 4 © Hodder & Stoughton Ltd 2013

3D shapes 2

Learning objectives

- Identify, describe, visualise, draw and make a wider range of 2D and 3D shapes including a range of quadrilaterals, the heptagon and tetrahedron; use pinboards to create a range of polygons. Use spotty paper to record results. (4Gs1)
- Visualise 3D objects from 2D nets and drawings and make nets of common solids. (4Gs4)

Resources

Sets of cards made from photocopiable pages 103 and 104 (from which you have removed the sphere and the hemisphere); sets of cards made from photocopiable pages 106 and 107; empty cardboard packaging in various simple shapes, e.g. cubes, cuboids, triangular prisms, cones and cylinders; thin card; rulers; scissors; sticky tape; art straws; modelling clay.

Starter

- Organise the learners into small groups of two to four. Give each group a set of cards made from photocopiable pages 103 and 104 (from which you have removed the sphere and the hemisphere) and a set of cards made from photocopiable pages 106 and 107.
- Ask the learners to match each net to the 3D shape it makes. Provide an extra challenge by giving a time limit, or by asking groups to race against each other to be the first to match all the shapes correctly.

Main activities

- Demonstrate identifying the net of a simple 3D shape, for example a cuboid, triangular prism or cylinder, by unfolding cardboard packaging. Demonstrate how to use the unfolded packaging as a model for drawing a net that will allow you to construct a similar shape of a different size, for example half the size, twice the size or 2 cm longer along each edge.

- Give out the cardboard packaging, thin card, rulers, scissors and sticky tape. Ask each learner to choose a package, unfold it, and use it as a model for constructing a similar card shape in a different size.
- Demonstrate using art straws joined with modelling clay how to make a skeleton shape of the same dimensions as one of the shapes made from card. Give out the art straws and modelling clay and ask the learners to make a skeleton model of their card shape.

Plenary

- Ask the learners what they found most difficult about making their shapes.
- Describe the net of a 3D shape. Ask the learners to visualise the net from your description and then visualise the 3D shape it makes. Ask: *What shape does the net I described make?*

Success criteria

Ask the learners:

- Can you match this net with the 3D shape it makes?
- Can you match this 3D shape with the net that makes it?
- Show me a 3D shape you constructed. Can you explain how you made it?
- Choose one of the shapes you made. Can you draw a sketch of it?

Ideas for differentiation

Support: Group these learners together and work with them, helping them to draw nets and construct shapes.

Extension: Challenge these learners to construct card and art straw models of various types of pyramid.

Nets of 3D shapes 1

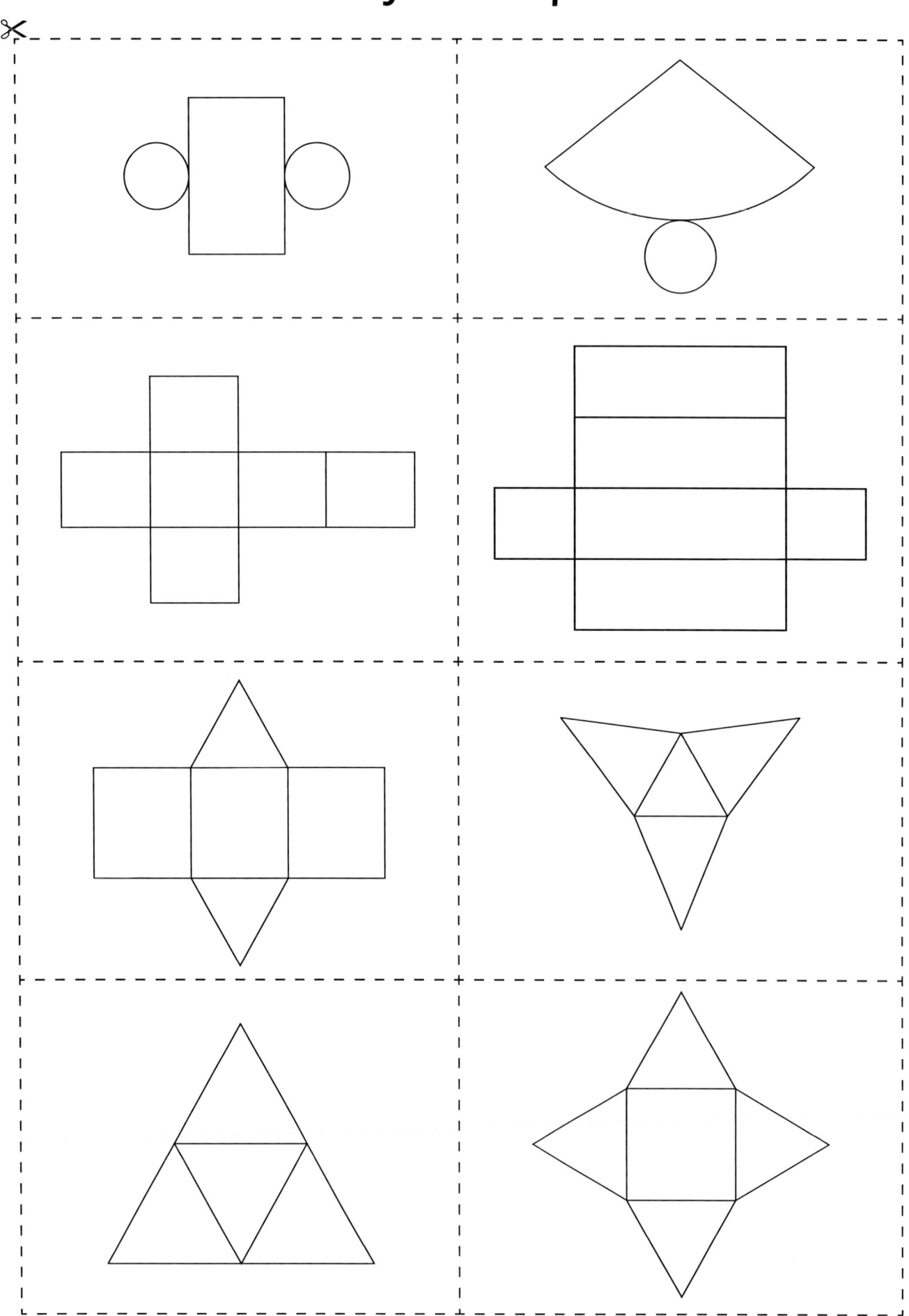

Cambridge Primary: Ready to Go Lessons for Maths Stage 4 © Hodder & Stoughton Ltd 2013

Nets of 3D shapes 2

Symmetry

Learning objectives

- Identify and sketch lines of symmetry in 2D shapes and patterns. (4Gs3)
- Find examples of shapes and symmetry in the environment and in art. (4Gs5)
- Identify simple relationships between shapes. (4Ps7)

Resources

Photocopiable page 109; objects decorated with symmetrical patterns, e.g. wallpaper, wrapping paper, fabric, tiles; large collection of small, coloured 2D shapes.

Starter

- Ask the learners: *What do we mean if we describe a shape as symmetrical?* (We mean that it has at least one line of symmetry. A line of symmetry is an imaginary line through the centre of the shape along which you could fold it and both halves would match exactly.) Ask the learners to look around the classroom and make a list of all the symmetrical shapes they can see.
- For each object identified as symmetrical, ask the learners to say how many lines of symmetry it has.

Main activities

- Hand out photocopiable page 109. Discuss the symmetrical patterns shown in the illustrations. Ask: *How many lines of symmetry does each pattern have?* Ask the learners to draw in the lines of symmetry. Show your collection of objects decorated in symmetrical patterns and identify the lines of symmetry in them.
- Give out the small, coloured 2D shapes. Ask the learners to make a symmetrical pattern by putting shapes together, taking colour into consideration as well as shape. Talk with the learners about their patterns, asking how many lines of symmetry it will have when it is finished.

- Ask the learners if they have used any regular polygons in their patterns (polygons whose sides are all equal and whose angles are all equal). Ask: *Which regular polygon has the fewest sides?* (The equilateral triangle.) *How many sides does it have?* (Three.) *How many lines of symmetry does it have?* (Three.)
- Ask: *Is it true that the number of lines of symmetry a regular polygon has is always equal to the number of its sides? Are there any exceptions? How do you know? What did you do to find out?*

Plenary

- Ask the learners to explain their conclusions from their investigation into the number of lines of symmetry in regular polygons.
- The learners should be able to confirm that the statement that the number of lines of symmetry a regular polygon has is always equal to the number of its sides is true.

Success criteria

Ask the learners:

- Can you name a symmetrical object in the classroom?
- How many lines of symmetry does it have?
- Can you show me where the lines of symmetry are on this shape / pattern?
- What can you tell me about the number of lines of symmetry in a regular polygon?

Ideas for differentiation

Support: Provide these learners with support in the final Main activity by providing them with regular polygons to draw around.

Extension: Ask these learners to investigate the number of lines of symmetry in various types of parallelogram.

Symmetrical patterns in art

Rangoli design

Celtic design

Islamic design

Mandala design

Navajo design

Aboriginal design

Co-ordinates and position

Learning objectives

- Describe and identify the position of a square on a grid of squares where rows and columns are numbered and / or lettered. (4Gp1)
- Devise the directions to give to follow a given path. (4Gp3)

Resources

Masking tape; a 'prize' (e.g. a picture of a ripe fruit); photocopiable page 111; squared paper.

Starter

- Before the lesson, use masking tape to make a 10 × 10 square grid on the floor of the classroom or hall. Each square should be big enough for a learner to stand inside. Mark a random selection of squares in the grid with a masking tape 'X'.

- Place the 'prize' in one of the corner squares, and ask a learner to stand in the opposite corner. Explain that the squares with the crosses are trees and must be moved around. Ask the rest of the learners to give the learner in the grid directions to get to the prize, for example: *Walk two squares forward. Make a quarter turn left. Walk three squares forward. Make a quarter turn right.*

Main activities

- Hand out photocopiable page 111. Ask the learners to give instructions that will allow the rocket to reach the moon without hitting an obstacle. Ask: *How many instructions did you give? What is the fewest number of instructions you need?*

- Revise how co-ordinates work (across then up) and how they are written. Ask the learners questions about the co-ordinates of the objects on photocopiable page 111, for example: *What's in square (5,6)? Write down the co-ordinates of the square the alien astronaut is in.*

- Give out squared paper and ask the learners to devise their own co-ordinates grid on which someone or something is trying to get somewhere, but there are obstacles in the way. Ask them to offer ideas for the context (for example a child walking to school, a mouse going after a piece of cheese, or a treasure-hunter avoiding traps). Ask the learners to write questions about their grid for a friend to answer.

Plenary

- Ask selected learners to show the co-ordinate grids they created.

- Ask the learners questions with the following format: *Start on (x, y). Move 2 / 3 / 4 squares right / left and 2 / 3 / 4 squares up / down. Which square do you end on?*

Success criteria

Ask the learners:

- Can you give directions for the alien spaceship to get to the moon?
- Can you point to square (3,6)? What about (6,3)?
- What is in square (5,3)?
- What are the co-ordinates of the square the ringed planet is in?

Ideas for differentiation

Support: In the final Main activity, help these learners to draw a 10 × 10 grid in which the horizontal axis is lettered rather than numbered, producing co-ordinates such as J5 or D3.

Extension: Ask these learners to write more questions in the final Main activity.

Rocket to the moon

Give instructions so that the rocket reaches the moon without hitting anything!

Angles

Learning objectives

● Know that angles are measured in degrees and that one whole turn is 360° or four right angles; compare and order angles less than 180°. (4Gp2)

Resources

Photocopiable page 113; right-angle measurers; straight-line angle measurers.

Starter

• On the board, draw a pair of angles, for example:

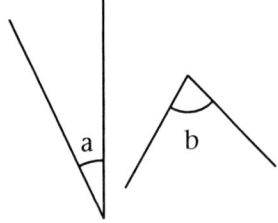

• Ask the learners to say which angle is larger and explain why (angle b because the amount of turn between its arms is greater). Repeat for other pairs of angles (all less than 180°).

• On the board draw about half a dozen angles of varying sizes, all less than 180°. Give each angle a unique letter or number. Ask the learners to list the angles in order from smallest to largest. Repeat for other sets of angles.

Main activities

• Introduce the degree (°) as the unit used to measure an amount of turn (and so the size of angles). Explain that there are 360° in a complete turn.

• Revise the fact that there are four right angles in a complete turn. Ask the learners how many degrees there must be in a right angle (90), and how many in a straight line angle (180).

• Copy the table from photocopiable page 113 onto the board. Draw a few angles on the board, giving each a letter. Ask for volunteers to write each angle's letter in the correct column of the table.

• Give each learner photocopiable page 113 and ask them to write each angle's letter in the correct column in the table. Ask finishers to draw three more angles for each column of the table.

Plenary

• Draw an angle that can be compared easily to a right angle, for example 45° (half a right angle); 30° (one-third of a right angle), 60° (two-thirds of a right angle) or 135° (one and a half right angles).

• Ask the learners to compare the angle to a right angle, and then estimate the number of degrees in the angle.

Success criteria

Ask the learners:

● What unit is used to measure angles?

● How many degrees are there in one complete turn?

● How many degrees are there in a right angle?

● (Pointing to an angle on photocopiable page 113:) Which column in the table does this angle belong in? Why?

Ideas for differentiation

Support: In the final Main activity, give these learners right-angle measurers (quarter-circles) and straight-line angle measurers (semi-circles).

Extension: Ask these learners to classify angles greater than 180° but less than 360°.

Name: _____

How many degrees?

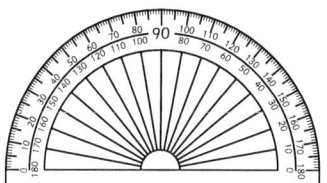

Write the letter of each angle in the correct column of the table.

Less than 90°	90°	Greater than 90° but less than 180°	180°

a)

b)

c)

d)

e)

f)

g)

h)

i)

j)

k)

l)

m)

n)

o)

p)

Unit assessment

Questions to ask

- (Pointing to a selection of 3D shapes:) Which of these shapes are pyramids?
- How are all pyramids the same?
- Which of the pyramids is a tetrahedron? What is special about it?

- Choose and name two 2D shapes. What is the same about them? What is different about them?
- Can you draw in the lines of symmetry on this shape / pattern?

Summative assessment activities

Observe the learners while they take part in these activities. You will quickly be able to identify those who appear to be confident and those who may need additional support.

Polygon sort

This activity assesses the learners' ability to classify polygons using criteria such as the number of right angles, whether or not they are regular and their symmetrical properties.

You will need:

A selection of polygons, e.g. various types of triangle including scalene, isosceles, right angle and equilateral; various types of quadrilateral including square, rectangle, rhombus, parallelogram and kite; regular and irregular pentagons, hexagons, heptagons and octagons.

What to do

- Ask the learners to sort the polygons into two groups: regular and irregular. A shape that might catch the learners out is the rhombus (it is irregular because its angles are not all equal).

- Ask the learners to sort the polygons into groups according to the number of right angles they have, and then according to the number of lines of symmetry they have (allow the learners to devise their own groupings within these criteria).

- Ask them to sort the polygons in any way they choose, and explain how they have sorted them.

Just triangles

This activity assesses the learners' ability to describe, visualise and make a wider range of 3D shapes.

You will need:

A tetrahedron; lots of identical equilateral triangles (ideally that click together).

What to do

- Display the tetrahedron and ask the learners to describe as many of its properties as they can.

- Ask them to make a tetrahedron from equilateral triangles.

- Ask: *Can you use just equilateral triangles to make any other closed 3D shapes?* If the learners need encouragement, explain that there are two more closed 3D shapes that can be made just from equilateral triangles, and that they both use more triangles than the tetrahedron. You might further suggest that the learners first try to make a shape out of five triangles, then six, seven, and so on.

- Ask the learners to describe the shapes they've made, and compare them with each other and with the tetrahedron. Introduce the shapes' names (the 8-sided shape is an octahedron and the 20-sided shape is an icosahedron).

Written assessment

Distribute photocopiable page 115. Ask the learners to read the questions and write the answers. They should work independently.

Name: _____

Looking at shapes

1. Name these shapes.

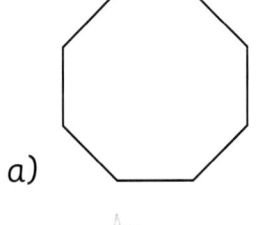

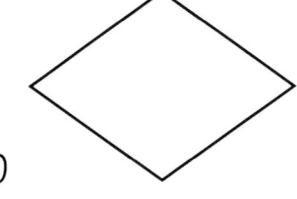

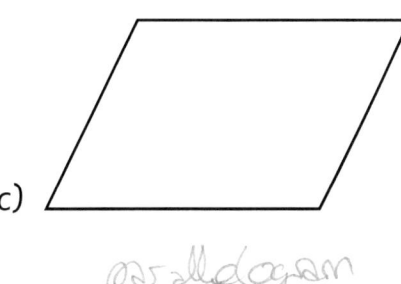

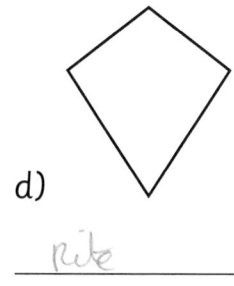

a) _octagon_   b) _rhombus_   c) _parallelogram_   d) _kite_

2. (Circle) the odd shape out in the set. Explain why it is different.

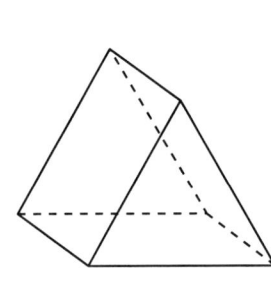
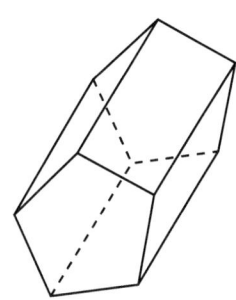
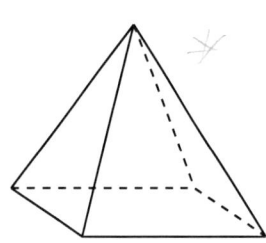
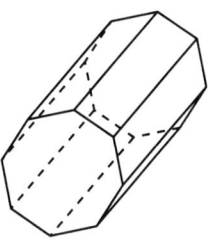

it is not a prism

3. Sketch in the lines of symmetry on these shapes.

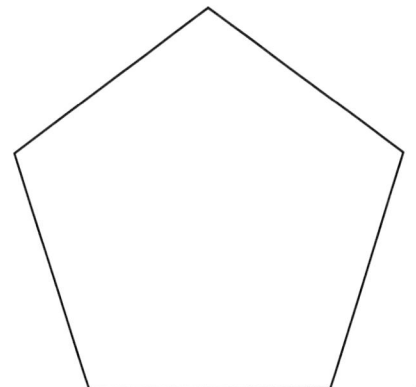
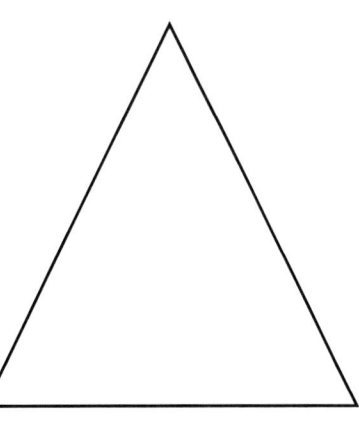
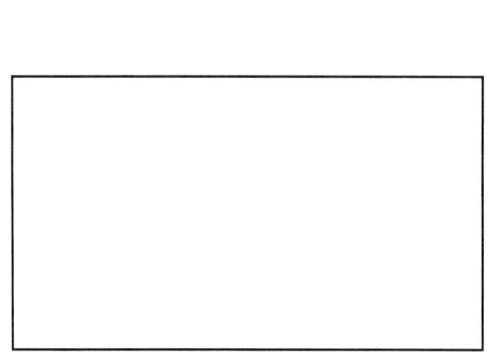

4. On centimetre-squared paper, draw a 10 square by 10 square grid. Label the rows and columns from 0 to 9. Colour in the following squares on the grid: (3,4), (7,5), (1,0), (6,8), (2,9).

5. On the back of this page, draw three angles less than 90° and three angles between 90° and 180°.

Mass 2

Learning objectives

- Choose and use standard metric units and their abbreviations when estimating, measuring and recording length, weight and capacity. (4Ml1)
- Where appropriate, use decimal notation to record measurements, e.g. 1.3 m, 0.6 kg, 1.2 l. (4Ml3)
- Interpret intervals / divisions on partially numbered scales; record readings accurately. (4Ml4)

Resources

Cards made from photocopiable page 117; multiple sets of six sand bags (sealed plastic bags containing sand) with the following masses: A: 2.1 kg, B: 0.7 kg, C: 2.8 kg, D: 1.5 kg, E: 1.2 kg and F: 0.5 kg; standard 1 kg masses; a range of mass measuring instruments suitable for measuring masses up to 3 kg (e.g. pan balances, spring scales, kitchen scales).

Starter

- Organise them into pairs and give each pair a set of cards made from photocopiable page 117.
- Ask them to match sets of cards showing the same mass.
- You could increase the challenge by giving the learners a fixed time limit, or challenging pairs to be the first to finish matching all their cards correctly.
- Go through the answers together.

Main activities

- Organise the learners into groups of three or four and give each group a set of sand bags and two standard 1 kg masses.
- Draw the following table on the board and ask the learners to copy it:

Sand bag	A	B	C	D	E	F
Estimated mass						
Measured mass						

- Ask the learners to estimate the mass of each bag, and record their estimates in kg, using decimal notation, for example 1.6 kg.
- Ask the learners to share their estimates, and describe the strategies they used in order to make them (for example ordering the bags from lightest to heaviest, and / or making direct comparisons with the 1 kg masses).
- Make the mass measuring instruments available and ask the learners to use them to find and record the actual mass of each bag.

Plenary

- Reveal the correct mass of each bag. Ask the learners to mark each other's estimates and measurements, awarding the following points:
  - Estimates: 3 points within 0.1 kg; 2 points within 0.3 kg; 1 point within 0.5 kg
  - Measurements: 3 points for each correct measurement.

Success criteria

Ask the learners:

- Estimate the mass of this bag. How did you make your estimate?
- How would you write 2 kg 600 g using just kilograms?
- What is the reading on this scale?
- How would you write it down?

Ideas for differentiation

Support: In the Main activity, put these learners in a group with more confident learners who work well with others.

Extension: Ask these learners, once they have finished the Main activity, to make up their own activity based on estimating and measuring masses.

Mass matching cards

	600 g	0.6 kg
(number line: 400 g, 600 g, 800 g)		
(number line: 1 kg, 800 g, 600 g)	900 g	0.9 kg
(number line: 0.5 kg, 1 kg, 1.5 kg)	1 kg 300 g	1.3 kg
(number line: 1.9 kg, 1.8 kg, 1.7 kg, 1.6 kg, 1.5 kg)	1600 g	1 kg 600 g
(number line: 1700 g, 1800 g, 1900 g, 2000 g)	1 kg 800 g	1.8 kg
(number line: 2 kg, 1.8 kg, 1.6 kg, 1.4 kg)	1900 g	1 kg 900 g

Capacity 2

Learning objectives

- Understand everyday systems of measurement in length, weight, capacity and time, and use these to solve problems as appropriate. (4Pt2)
- Estimate and approximate when calculating, and check working. (4Pt8)
- Make up a number story for a calculation, including in the context of measures. (4Ps1)
- Explain methods and reasoning orally and in writing; make hypotheses and test them out. (4Ps9)

Resources

Teaspoon; litre jug; about half a dozen containers with capacities less than 1 litre, each of which has a different capacity (e.g. ladle, teacup, mug, shampoo bottle, handwash bottle, cordial bottle); photocopiable page 119.

Starter

- Before the lesson, find and record the capacity of each container you are using.
- Display a teaspoon and a litre jug and ask the learners to give the capacity of each. Label the teaspoon 5 ml and the litre jug 1 l or 1000 ml.
- Display the collection of containers. Ask the learners to order the containers from least to greatest capacity, and then estimate the capacity of each container in millilitres, and record their estimates.
- Reveal the actual capacity of each container. The learners score a point for every estimate within 50 ml of the correct capacity.

Main activities

- Display a copy of photocopiable page 119 and read out one of the word problems. Ask: *What calculation do you need to do in order to solve the problem?*
- Ask the learners to estimate the answer, and describe the strategies they used (for example rounding to the nearest 10 or 100).

- Work through the calculation, asking the learners to suggest the method.
- Ask the learners to compare the answer with the estimate. Ask: *Given the estimate, does the answer seem reasonable? What would it mean if the estimate and the answer were very different?* (It might mean the answer is wrong.) *What else could you do to check the answer?* (You could repeat the calculation using a different method, or use the inverse operation.)
- Distribute photocopiable page 119. Ask the learners to complete it, working either individually or in pairs. Remind them to make an estimate for each question and use it to check the reasonableness of their answer.

Plenary

- Ask the learners to give the answers to the word problems on photocopiable page 119, describing the methods they used, and suggesting a calculation that can be used to check each answer.
- Ask volunteers to read out the number stories they wrote.

Success criteria

Ask the learners:

- Estimate the capacity of this container. How did you work out your estimate?
- Choose one of the problems on photocopiable page 119 that you have already answered. How did you work out the answer?
- How could you check your answer is correct?
- Can you make up a number story to go with this calculation: 6 × 250 ml = 1.5 litres?

Ideas for differentiation

Support: Group these learners together and guide them through an extra problem. Ask them to miss out questions 5 to 7.

Extension: Ask these learners to make up word problems based on their own ideas instead of the questions given on photocopiable page 119.

Name: _____

Capacity problems

1. A recipe uses 270 ml of water and 850 ml of stock. How much is this in total?

2. A tap drips for four hours. Every hour, 750 ml of water drips from the tap. How much water does the dripping tap waste?

3. A full bottle of cooking oil contains 1 litre 250 ml. 800 ml of oil is left in the bottle. How much oil has been used?

4. Three friends had a drink of milk. One friend drank 330 ml, one drank 275 ml and one drank 410 ml. How much milk did they drink altogether?

5. A water butt contains 58 litres of water.

   a) How many 5 litre buckets could you fill from the water butt?

   b) How many litres of water would be left in the water butt?

6. A car's petrol tank holds 56 litres when it is full. At the moment, the petrol tank is $\frac{1}{4}$ full. How many litres of petrol are in the tank?

7. Write a number story on the back of this page to go with each of these calculations:

   a) 3×700 ml

   b) $1\,l \div 4$

   c) 350 ml $- 125$ ml

Length 2

Learning objectives

- Choose and use standard metric units and their abbreviations when estimating, measuring and recording length, weight and capacity. (4Ml1)
- Know and use the relationships between familiar units of length, mass and capacity, know the meaning of kilo-, cent-, and milli-. (4Ml2)
- Where appropriate, use decimal notation to record measurements, e.g. 1.3 m, 0.6 kg, 1.2 l. (4Ml3)

Resources

Cards made from photocopiable page 121; unmarked sticks exactly 1 metre long; metre sticks; tape measures.

Starter

- Organise the learners into groups of two to four. Give each group a set of cards made from photocopiable page 121.
- Ask the learners to match sets of equivalent lengths, and then order the matched lengths from shortest to longest. Challenge them to match and order the cards within a time limit, or challenge groups to compete with each other to be the first to finish.

Main activities

- Call up three volunteers of varying heights. Hold up an unmarked metre-long stick against each volunteer and ask the rest of the class to estimate each volunteer's height in metres, using decimal notation. Record the estimates.
- Ask two of the volunteers to measure the other volunteer's height, using their own choice of equipment and method. Ask: *Do you think their measurement will be accurate? Why? Why not?* Challenge the learners to refine and improve the techniques. Record the measurements.

- Ask the learners some word problems about the volunteers' heights, for example:
  - *A doorway is 1.9 m high. How much headroom will Volunteer 1 have when he walks through it?*
  - *I'm 26 cm taller than Volunteer 2. How tall am I?*
  - *Volunteer 3 has grown 7 cm in the past year. How tall was he one year ago?*
- Organise the learners into groups of three, and ask them first to estimate and then to measure and record their heights.
- Ask each learner to write a word problem about their own height. It does not need to be true! Ask each group to give their problems to another group to solve.

Plenary

- Ask the learners whether their estimates got closer to the actual measurements as the lesson progressed. If so, ask them to suggest why.
- Ask a few learners to share the word problems they have written and invite the rest of the class to solve them.

Success criteria

Ask the learners:

- How do you write 123 cm in metres?
- (Pointing to a learner in a different group:) About how tall do you think this person is? How did you make your estimate?
- Can you measure their height? How close was your estimate?
- Can you write their height in metres?

Ideas for differentiation

Support: Group these learners together, and assist them with making and recording their estimates and measurements.

Extension: Ask these learners to write more than one word problem each.

Equivalent length cards

1 m 1 cm	101 cm	1.01 m
1 m 5 cm	105 cm	1.05 m
1 m 10 cm	110 cm	1.1 m
1 m 15 cm	115 cm	1.15 m
1 m 20 cm	120 cm	1.2 m
1 m 25 cm	125 cm	1.25 m
1 m 50 cm	150 cm	1.5 m
1 m 75 cm	175 cm	1.75 m
1 m 90 cm	190 cm	1.9 m

Area and perimeter 2

Learning objectives

● Understand that area is measured in square units, e.g. cm squared. (4Ma2)

● Find the area of rectilinear shapes drawn on a square grid by counting squares. (4Ma3)

Resources

Photocopiable pages 123 and 124; scissors; newspaper; masking tape; marker pens.

Starter

- Organise the learners into pairs and give each pair photocopiable pages 123 and 124 and a pair of scissors.
- Ask the pairs to cut out each shape and put them in order according to their area.
- Discuss answers. Ask the learners how they decided which shapes were larger or smaller (by counting the squares).

Main activities

- Revise the square centimetre as a unit of area. Relate it back to the Starter activity, in which the squares on the grid were square centimetres. Introduce the square metre (a square one metre along each side) as the next largest unit of area.
- Organise the learners into pairs, giving each pair several sheets of newspaper, a pair of scissors, masking tape and a marker pen. Ask each pair to make a square metre out of newspaper and label it '1 square metre'.
- Ask the learners to estimate the area of various objects in the classroom in square metres, for example the area of a window, the floor area covered by a cupboard or a rug, the area of a table top.
- Present the following problem: *You have a horse you want to put in a square or rectangular enclosure. You have 360 metres of fencing. What is the largest enclosure you can make?*

Plenary

- Ask the learners to give their solutions to the word problem, and explain their method.
- Reveal the correct answer to the word problem: a square enclosure 90 m along each side gives the largest area (8100 square metres).

Success criteria

Ask the learners:

● (Pointing to a shape from photocopiable pages 123 and 124:) What is the area of this shape? How do you know?

● Imagine a rectangle that is 20 m long and 70 m wide. What is its area? How did you work it out?

● Imagine a rectangle that is 50 m long and 40 m wide. What is its perimeter? How did you work it out?

Ideas for differentiation

Support: Reframe the word problem like this for these learners: *You have a rabbit you want to put in a square or rectangular enclosure. You have 36 metres of wire mesh. What is the largest enclosure you can make?*

Extension: Ask these learners who have solved the word problem to investigate whether, for a given perimeter, a square will always give a greater area than a rectangle.

Ordering areas 1

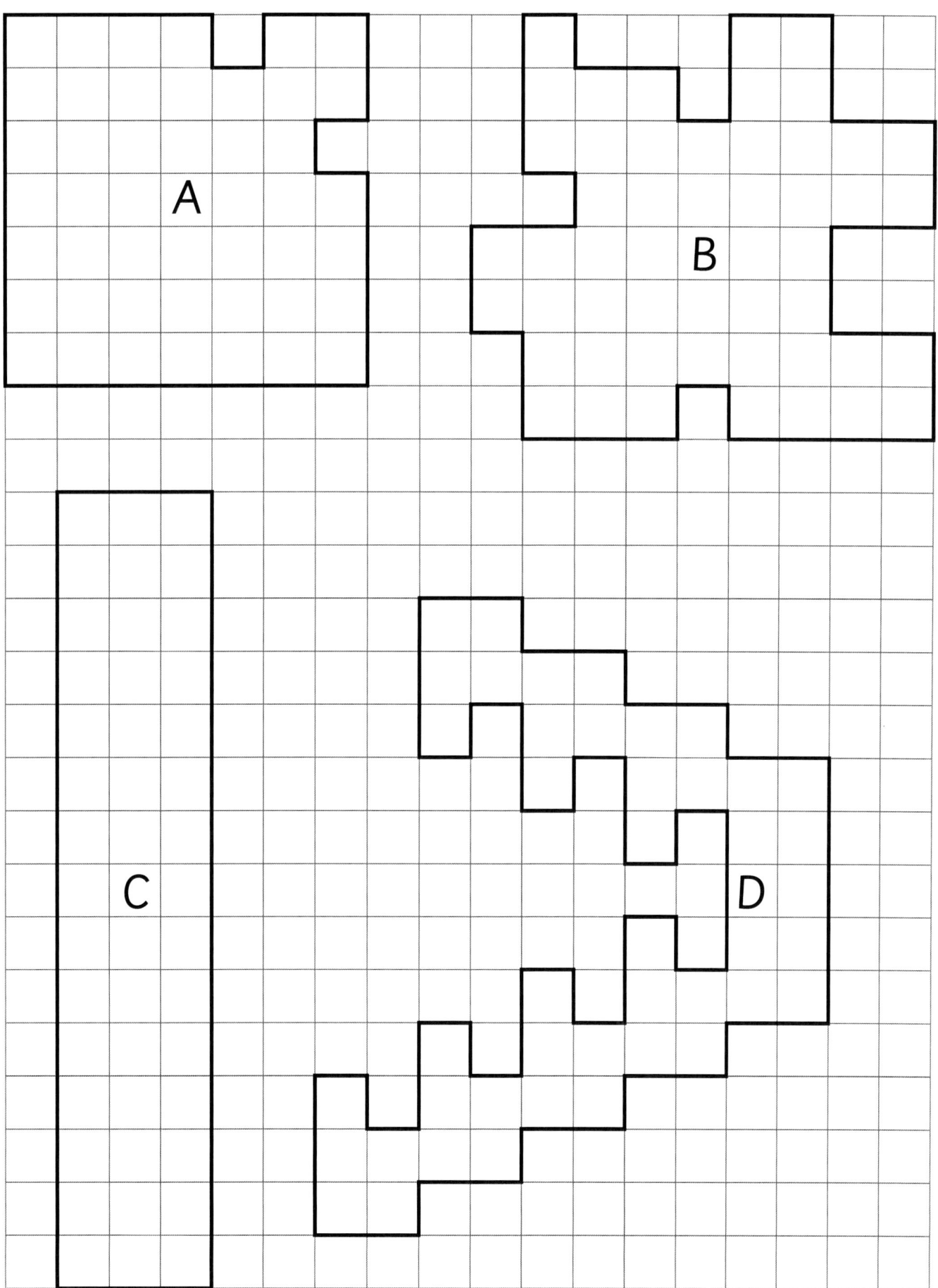

Ordering areas 2

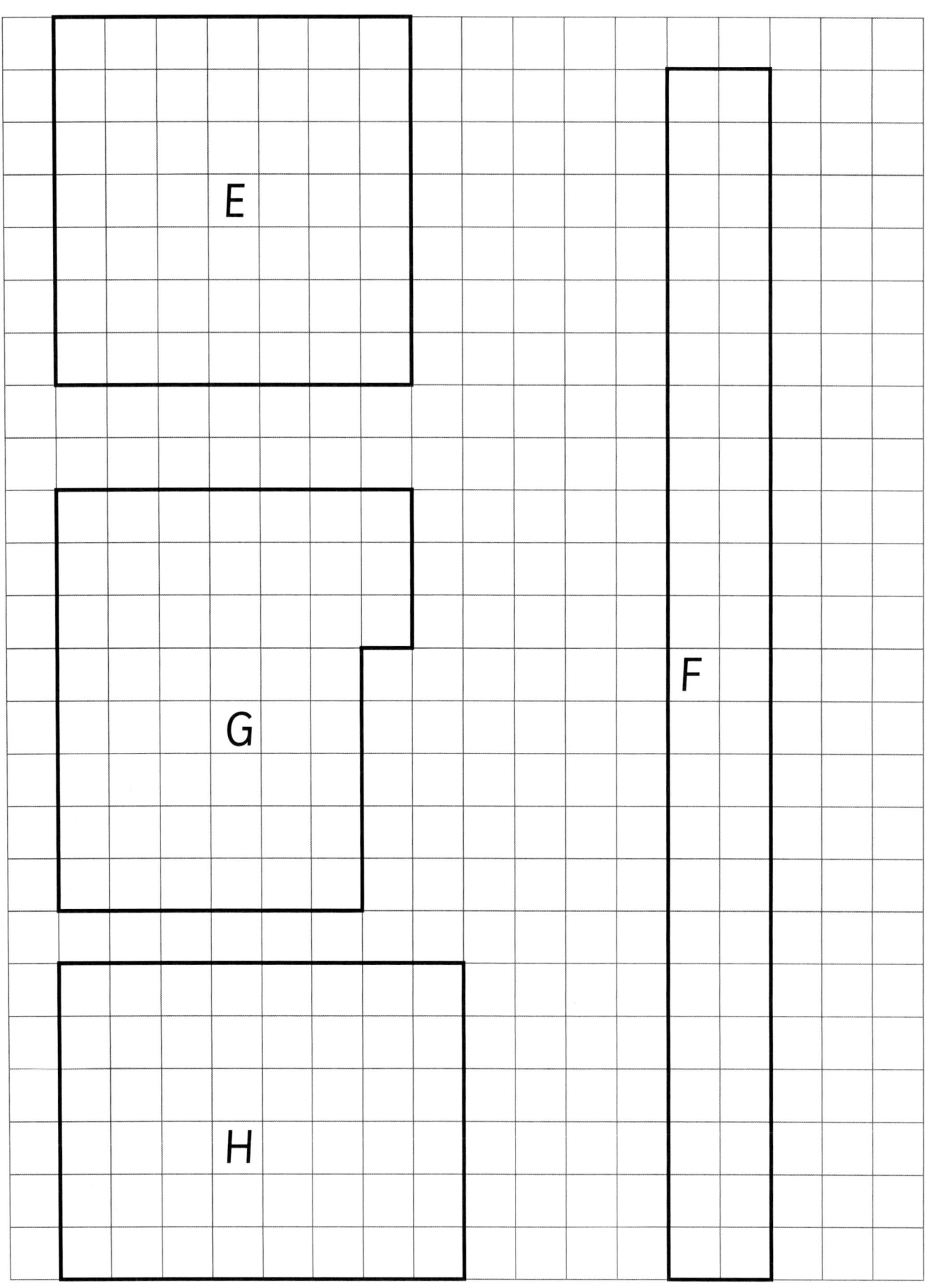

 Cambridge Primary: Ready to Go Lessons for Maths Stage 4 © Hodder & Stoughton Ltd 2013

Time 4

Learning objectives

● Read and tell the time to the nearest minute on 12-hour digital and analogue clocks. (4Mt1)

● Use am, pm and 12-hour digital clock notation. (4Mt2)

● Understand everyday systems of measurement in length, weight, capacity and time, and use these to solve problems as appropriate. (4Pt2)

Resources

Cards made from photocopiable pages 47 and 48; one large and lots of small clock faces with movable hands; photocopiable page 126.

Starter

• Organise the learners into groups of four. Give each group a set of cards from photocopiable pages 47 and 48.

• Ask the learners to separate out the cards showing times in words, shuffle them and put them in a face-down pile. Shuffle together the clock cards, and deal out six cards to each player.

• Turn over the top card from the pile. The first learner to shout 'Snap!' and show a matching card from their hand 'captures' both cards. The winner is the player with the most captured cards when all the cards in the pile have been turned over.

Main activities

• Give out the small clock faces. Give the learners practice in making times to the nearest minute (given in writing, in digital format or verbally) on the clock faces. Confirm correct answers using the large clock face.

• Use word problems based around the school day to give practice and consolidation in telling the time, for example: *In forty-five minutes it's break time. What time will it be then? Break lasts 15 minutes. What time does it finish? The lesson after break lasts 1 hour 15 minutes. What time does it finish?* For each question, ask the learners to make the time on their clock faces. Confirm correct answers using the large clock face.

• Hand out photocopiable page 126. Read through the first question together with the learners, drawing in the answer on the blank clock face. Ask the learners to complete the questions on the photocopiable page.

Plenary

• Confirm the correct answers to the word problems on photocopiable page 126 by asking the learners to make the answers on the large clock face.

• Ask volunteers to read out the word problems they wrote. Ask the other learners to solve one or more of the problems.

Success criteria

Ask the learners:

● What time does this digital clock show?

● What time does this analogue clock show?

● Can you make 6:54 on an analogue clock?

● What abbreviation would you put after 6:54 to show it's a morning time?

Ideas for differentiation

Support: Group these learners together and guide them through an extra problem (Question 2). Ask them to attempt only the even numbered questions.

Extension: Challenge these learners to write and solve word problems involving time intervals that are not multiples of 5 minutes.

Name: _____

Time problems 2

1. A flight to Bahrain was supposed to depart
    at 10:55, but it was delayed by 20 minutes.
    At what time did it depart? 11:15

2. A flight to Cairo was supposed to depart
    at 3:25, but it was delayed by 45 minutes.
    At what time did it depart? 4:10

3. A flight from Amman was supposed to arrive
    at 12:40, but it was delayed by 35 minutes.
    At what time did it arrive? 1.15

4. A flight from Doha was supposed to arrive
    at 1:20, but it was delayed by 55 minutes.
    At what time did it arrive? 2.15

5. The Khan family were supposed to check in for
    their flight to Dubai at 6:30, but they were
    40 minutes late. At what time did they check in? 7.10

6. The Nasser family's flight from Jeddah arrived
    at 12:50. It took them 25 minutes to collect
    their baggage and clear customs. At what
    time were they ready to leave the airport? 1.15

7. Make up some more time problems that can
    be answered by drawing a time on a clock
    face. Write them on the back of this page.

Cambridge Primary: Ready to Go Lessons for Maths Stage 4 © Hodder & Stoughton Ltd 2013

Time 5

Learning objectives

- Understand everyday systems of measurement in length, weight, capacity and time, and use these to solve problems as appropriate. (4Pt2)
- Choose units of time to measure time intervals. (4Mt4)

Resources

Sticky notes; cards made from photocopiable pages 128 and 129; A3 paper; glue sticks.

Starter

- Organise the learners into pairs, asking each pair to write down as many units of time as they know. Ask them to write each unit of time on a separate sticky note.
- Ask them to order the units of time from shortest to longest.
- Ask them to annotate each sticky note to show the relationship between that unit of time and another unit of time, for example 1 day = 24 hours.
- Discuss answers, revising the following units of time and the relationships between them: second, minute, hour, day, week, month, year, decade (10 years), century (100 years), and millennium (1000 years).

Main activities

- Organise the learners into pairs and give each pair a set of cards made from pages 128 and 129, a piece of A3 paper, a glue stick and ten sticky notes. Ask the learners to sort the events on the cards into groups, depending on the unit of time used to measure them.
- When the learners are happy with their groupings, ask them to record them by sticking the grouped cards onto the A3 paper.
- Ask them to choose ten more activities that are measured using a variety of units of time. Ask them to write each activity on a sticky note and give the sticky notes to another pair to sort and add to their piece of paper.

Plenary

- Call out a unit of time. Ask the learners to give an example of an event from the cards measured in that unit (or a combination of units that includes that unit), plus one of their own suggestions.
- Ask: *Were there any events that were difficult to sort? Which ones? Why?*

Success criteria

Ask the learners:

- What unit or units of time would you use to measure how long the dry season lasts?
- What is the name given to a period of 100 years?
- How many minutes are there in an hour?
- Name an event you might measure in minutes and seconds.

Ideas for differentiation

Support: In the first Main activity, provide these learners with a sorting table with seven columns labelled: seconds, minutes and seconds, hours and minutes, days, weeks, months, and years.

Extension: Ask these learners to extend their sorting table to include decades, centuries and millennia. Ask them to suggest events that belong in each of these regions of the table.

How long does it take? cards 1

running a sprint race	pouring a glass of water
writing your name	dialling a phone
calling the register	travelling to school
singing a song	getting dressed
watching a football match	reading a book
taking a flight	watching a film
having a cold	drying dates

 Cambridge Primary: Ready to Go Lessons for Maths Stage 4 © Hodder & Stoughton Ltd 2013

June						
M	T	W	Th	F	Sa	Su
	1	2	3	4	5	6
7	8	9	10	11	12	13
14	15	16	17	18	19	20
21	22	23	24	25	26	27
28	29	30				

celebrating a wedding	decorating a room
taking a holiday	a child becoming an adult
Ramadan	the season for fresh figs
learning Mathematics	the rainy season
the football season	growing rice
learning to play a musical instrument	growing an olive tree from a seed

Time 6

- Read simple timetables and use a calendar. (4Mt3)
- Understand everyday systems of measurement in length, weight, capacity and time, and use these to solve problems as appropriate. (4Pt2)
- Explain methods and reasoning orally and in writing; make hypotheses and test them out. (4Ps9)

Resources

Copies of a local bus or train timetable; copies of a calendar showing the whole of the current year; photocopiable page 131.

Starter

- Organise the learners into pairs, giving each pair a copy of a local bus or train timetable.
- Ask the learners questions whose answers can be found by reading and interpreting the timetable, for example ask for:
  - the time a particular bus / train arrives at or departs a given stop / station
  - the duration of a particular journey
  - how long you would have to wait until the next bus / train if one were cancelled
  - the time a bus / train will arrive at a given stop / station if it is running a particular number of minutes late
  - the departure time of the train or bus that gives the quickest journey time between two places.

Main activities

- Give out copies of the current year's calendar. Ask the learners to calculate, for example: the length of time since / until a given date; the length of time between two given dates, or the date of an event given another date (either earlier or later) and a duration of time expressed in days, weeks or a combination of the two. For each question, ask the learners to explain how they worked out the answer.

- Hand out photocopiable page 131. Ask the learners to use the current year's calendar to answer the questions on it.
- Discuss answers and methods.
- Challenge the learners to solve the following problem: 1 January 2018 will fall on a Monday. When will 1 January fall on a Monday again? Hint: Don't forget about leap years!

Plenary

- Ask the learners to give their solution to the problem, explaining their method and their reason for choosing it.
- Confirm that the next year in which 1 January will be a Monday is 2024 (2020 is a leap year, so has one extra day).

Success criteria

Ask the learners:

- If you wanted to travel from [Place A] to [Place B], and you had to arrive in [Place B] by [time], which bus would you get from [Place A]?
- How many buses to [Place C] leave [Place D] between 8a.m. and noon?
- Which month is it six months before April?
- What date is it three weeks and five days after 6 May?

Ideas for differentiation

Support: Give these learners an alternative problem in the final Main activity: Next year, 1 January will fall on a [day of the week]. On which days of the week will the first days of the rest of the months fall?

Extension: Challenge these learners to calculate the last year in which today's date fell on the same day of the week.

Calendar problems

1. Sara's birthday is on the third Thursday in August.
   What date is her birthday?

2. The Garcia family are going on a 14-night holiday.
   They leave on 9 June.
   What date do they get back home?

3. A certain crop takes five months to grow.
   A farmer plants the crop in November.
   When will the crop be ready to harvest?

4. Josh is taking part in a race on 15 December.
   He starts training 16 weeks before the race.
   On what date does he start his training?

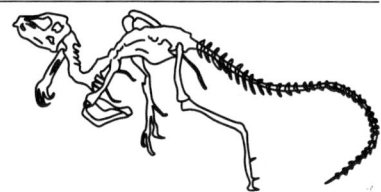

5. Entrance to a museum is free on
   the first Monday in every month.
   On what date in March is
   entrance to the museum free?

6. A school has a sports day on 21 July.
   The Year 4 trip takes place three weeks
   and four days before the sports day.
   What date does the Year 4 trip take place?

7. A theatre company is putting on a play.
   The first performance is on 10 May.
   They start rehearsals on 22 March.
   How long do the actors have to rehearse the play?

8. Write some more calendar problems of your own
   on the back of this page.

June						
M	T	W	Th	F	Sa	Su
	1	2	3	4	5	6
7	8	9	10	11	12	13
14	15	16	17	18	19	20
21	22	23	24	25	26	27
28	29	30				

Unit assessment

Questions to ask

- Which unit or units of time would you choose to measure how long a movie lasts?
- What is the time now? What will the time be in 45 minutes?
- Which unit would you use to measure the mass of this object? Can you estimate the mass of the object?
- Can you measure the mass of the object? How close was your estimate?
- Imagine an enclosure that is 20 m wide and 30 m long. What is its area? What is its perimeter?

Summative assessment activities

Observe the learners while they take part in these activities. You will quickly be able to identify those who appear to be confident and those who may need additional support.

Capacity estimating and measuring game

This game assesses the learners' ability to estimate, measure and record capacity.

You will need:

Range of waterproof containers; measuring jugs or cups marked in millilitres and litres; water.

What to do

- Organise the learners into groups of six. Give each group a range of waterproof containers and a selection of measuring jugs or cups.
- Each player should choose one waterproof container. Each player should then write the names of the six chosen objects in order, from smallest to greatest capacity, then estimate the capacity of each object in millilitres or litres and record their estimates on their ordered list.
- Each player must find and record the capacity of the container they chose, then compare each player's estimate with this actual capacity, giving one point to whichever player has the closest estimate. The player with the most points wins.

Length memory game

This game assesses the learners' knowledge of the relationships between familiar units of length.

You will need:

Cards made from photocopiable page 121.

What to do

- Before the lesson, sort the cards made from photocopiable page 121 into nine sets of three matched cards. Remove one whole set of matched cards and one card from each of the remaining sets, to leave eight pairs of matched cards.
- Organise the learners into groups of three. Give each group a set of prepared cards. Ask one learner to shuffle the cards and place them face down in a 4 × 4 grid.
- The players should take it in turns to turn over two cards. If the two cards show a matching length (for example 1 m 2 cm and 1.02 m), the player who turned them over on keep both cards. If the two cards do not match, the player should turn them back over.
- The game is over when there are no more cards left in the grid. The winner is the player with the most cards.

Written assessment

Distribute photocopiable page 133. Ask the learners to read the questions and write the answers. They should work independently.

Name: _____

All sorts of measures

1.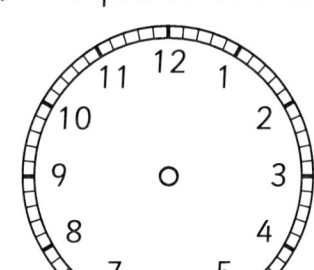

   500 g   700 g   900 g

   What is this measurement: a) in grams? _____

   b) in kilograms? _____

2. Complete the table below.

	Eli	Alfie	Sámi	Jake
Height in cm	130 cm			
Height in m		1.25 m		1.5 m
Height in m and cm			1 m 43 cm	

3. What units of time would you be most likely to use to measure the following events?

   a) a holiday _____

   b) a marathon _____

   c) the life of a person _____

4. A square walled garden has an area of 144 square metres.

   a) How long is the side of the garden? _____

   b) What is the perimeter of the garden? _____

5. Draw in the hands to show the correct time on each clock face.

   a) twenty-five past eight   b) a quarter to one   c) ten to seven

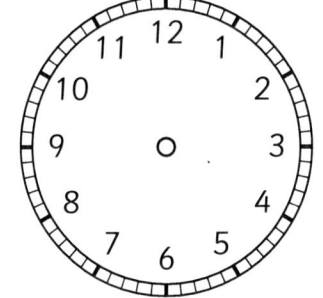
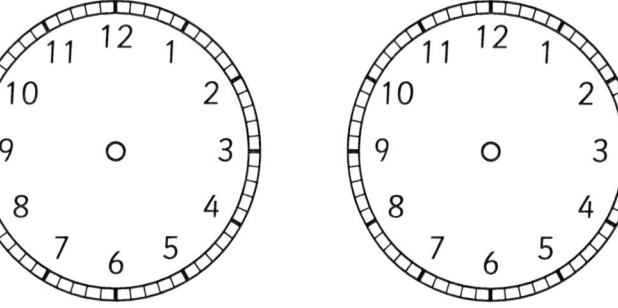

6. Express each of the times in Question 5 as digital times with a.m. or p.m.

Time a) is when Rhys goes to bed. _____

Time b) is when Rhys eats his lunch. _____

Time c) is when Rhys gets up. _____

Multiplying and dividing by 100

Learning objectives

● Multiply and divide three-digit numbers by 10 (whole number answers) and understand the effect; begin to multiply numbers by 100 and perform related divisions. (4Nn7)

Resources

Calculators; cards made from photocopiable page 135; place value charts.

Starter

- Hand out the calculators. Ask the learners to put a number into the calculator and then press × 100. Ask: *What do you notice? Try another number. Does the same thing happen?* Ask the learners to keep a written record of each calculation they do (for example 123 × 100 = 12 300).

- Ask the learners to describe a rule for multiplying by 100.

Main activities

- Draw on the board a place value chart like this:

Ten thousands	Thousands	Hundreds	Tens	Units

- Write a three-digit number on the place value chart and demonstrate multiplying it by 100, explaining why each digit moves two places to the left, and discussing the role of 0s as place holders. Give the learners plenty of practice in multiplying three-digit numbers by 100.

- Ask the learners to predict what will happen when you divide a number by 100 (the digits will move two places to the right). Using the place value chart, practise dividing four- and five-digit multiples of 100 by 100.

- Organise the learners into groups of four, giving each group a set of cards made from photocopiable page 135. Ask the learners to shuffle the cards and arrange them face down in a 6 × 6 grid. Ask them to take it in turns to turn over two cards. If the number on one card is 100 times the number on the other, the cards are a pair, and the learner who turned them over should keep them. If not, the learner must turn the cards back over. The game is over when there are no more cards left in the grid. The winner is the learner with the most cards.

Plenary

- Ask the learners to help you extend the place value chart two places to the right, to tenths (t) and hundredths (h). Write in a three-digit whole number (for example 394).

- Ask the learners to predict what will happen to the number when it is divided by 100.

Success criteria

Ask the learners:

● Can you explain the rule for multiplying numbers by 100?

● What is 504 × 100?

● Can you explain the rule for dividing numbers by 100?

● What is 38 100 ÷ 100?

Ideas for differentiation

Support: In the final Main activity, give these learners place value charts so that they can check the relationship between pairs of numbers by making them on a place value chart.

Extension: Challenge these learners to make their own sets of multiplying and dividing by 100 cards.

Multiplying and dividing by 100 cards

123	12 300	231	23 100
321	32 100	312	31 200
213	21 300	132	13 200
230	23 000	320	32 000
130	13 000	120	12 000
310	31 000	23	2300
32	3200	13	1300
12	1200	31	3100
210	21 000	21	2100

More multiplying and dividing by 100

Learning objectives

- Use decimal place value for tenths and hundredths in context, e.g. order amounts of money; convert a sum of money such as £13.25 to pence, or a length such as 125 cm to metres; round a sum of money to the nearest pound. (4Nn4)

Resources

A set of large cards made from photocopiable page 135; photocopiable page 137; place value charts (thousands, hundreds, tens, units, tenths, hundredths).

Starter

- Take a set of large cards made from photocopiable page 135, and sort the two- and three-digit numbers into one group and the four- and five-digit numbers into another group.

- Hold up one card at a time from the first group, and ask the learners to multiply the number by 100 and write down the answer. Ask them to say the number they have written.

- Hold up one card at a time from the second group, say the number aloud and ask the learners to divide the number by 100 and write down the answer.

Main activities

- Revise the relationship between dollars and cents. Establish that $1 = 100c$, so to convert dollars to cents you multiply by 100.

- Give the learners some simple conversions involving whole numbers of dollars, for example: *What is $6 converted to cents?* (600c.) *What is $15 converted to cents?* (1500c.)

- On the board, draw a Th, H, T, U, t, h place value chart. Write in an amount that is not a whole number of dollars, for example $8.95. Ask: *What is $8.95 converted to cents?* Demonstrate moving the digits two places to the left to find the answer (895 cents).

- Repeat the process for converting cents into dollars, establishing that $1c = \frac{1}{100}$ of $1, so to convert cents to dollars you divide by 100.

- Hand out the place value charts and photocopiable page 137. Read through the questions, ensuring the learners understand what they have to do. Ask them to work individually or in pairs to complete the photocopiable page.

Plenary

- Ask the learners to give their answers to the word problems on photocopiable page 137, explaining how they worked out each answer.

- Ask selected learners to read out word problems they have written themselves, and ask the rest of the class to solve them.

Success criteria

Ask the learners:

- Can you explain the rule for converting dollars to cents?
- What is $9.23 in cents?
- Can you explain the rule for converting cents to dollars?
- What is 412c in dollars?

Ideas for differentiation

Support: Ask these learners to attempt only the odd numbered questions on photocopiable page 137.

Extension: Challenge these learners to tackle photocopiable page 137 without using a place value chart.

Name: _____

Converting between dollars and cents

To convert dollars into cents you multiply by 100.

To convert cents into dollars you divide by 100.

1. You have been collecting 1c coins. You have collected 326 of them.

   How much money do you have in dollars? _____

2. Rashid has been given $2.70 to share with his two brothers, but he is finding it difficult to divide $2.70 by three. Rashid's father suggests converting the amount into cents first, and then dividing.

   a) How much does Rashid have in cents? _____

   b) How much money will each brother get? _____

3. At the bank I exchanged some 1c coins for $15.50.
   How many 1c coins did I exchange?

   _____

4. Fatima has eight 10c coins, six 5c coins and seven 1c coins in her purse. How much money does she have? Give your answer:

   a) in cents _____

   b) in dollars _____

5. Eight friends wrote down the amount of money in their pockets.
   The amounts were:

   $2.13   89c   195c   54c $0.85   208c $0.50 $1.87

   Order the amounts from smallest to largest.

   _____

6. Write some problems of your own that involve converting between dollars and cents. Give them to a friend to solve.

   _____

   _____

Negative numbers 2

Learning objectives

- Use negative numbers in context, e.g. temperature. (4Nn13)
- Make up a number story for a calculation. (4Ps1)

Resources

Counting stick; photocopiable page 139.

Starter

- Using the counting stick, lead the class in counting on and back in 1s, extending below 0.
- Repeat, counting in small whole-number steps, for example 2s, 3s, 4s or 5s, extending below 0 to at least −10.

Main activities

- Discuss various contexts in which negative numbers are used, for example to indicate temperatures below freezing, altitude below sea level or a withdrawal from a bank account.
- On the board, draw a number line from −10 to 10. Use the number line to perform simple additions that start on a negative number, for example −3 + 8; −9 + 5; −2 + 11. Do this by finding the start number on the number line and then counting on the number after the addition sign, for example to do −2 + 11, start at −2 and count on 11, to reach 9.
- Ask the learners to make up a number story for the calculation they have just done, for example: *Nina's bank account is $2 overdrawn. Nina deposits $11 into her bank account. The balance of her account is now $9.*
- Use the number line to perform simple subtractions that end on a negative number, for example 2 − 5; −1 − 3; 4 − 12. Do this by finding the start number on the number line and then counting back the number after the subtraction sign, for example to do 4 − 12, start at 4 and count back 12, to reach −8.

- Ask the learners to make up a number story for the calculation they have just done, for example: *The temperature in the fridge is 4°C. The temperature in the freezer compartment is 12°C lower. The temperature in the freezer compartment is −8°C.*
- Ask the learners to draw their own −10 to 10 number line. Distribute photocopiable page 139 and ask the learners to use their number line to work out the answers to the additions and subtractions on the page.

Plenary

- Go through the answers to the calculations on photocopiable page 139.
- Organise the learners into pairs, and ask each pair to choose a calculation from photocopiable page 139 and write a number story to go with it. Share number stories.

Success criteria

Ask the learners:

- Can you describe a situation in which you might need to use negative numbers?
- What is −8 + 3? Explain how you worked out the answer.
- What is 3 − 10? Explain how you worked out the answer.
- Can you make up a number story to go with the calculation 3 − 4 = −1?

Ideas for differentiation

Support: Group these learners together and work through the first few calculations on photocopiable page 139 with them.

Extension: Ask these learners to make up their own addition and subtraction calculations that either start or end on negative numbers, and give them to a friend to solve.

Name: _____

Calculations with negative numbers

Do these calculations by counting on or back on the number lines.

1. −2 + 5 = _____

−10   0   10

2. 2 − 6 = _____

−10   0   10

3. −8 + 5 = _____

−10   0   10

4. 8 − 10 = _____

−10   0   10

5. −3 + 7 = _____

−10   0   10

6. −5 − 2 = _____

−10   0   10

7. −7 + 6 = _____

−10   0   10

8. 4 − 9 = _____

−10   0   10

9. −8 + 10 = _____

−10   0   10

10. 6 − 12 = _____

−10   0   10

Odds and evens 2

Learning objectives

- Recognise odd and even numbers. (4Nn15)
- Make general statements about the sums and differences of odd and even numbers. (4Nn16)
- Explore and solve number problems and puzzles. (4Ps4)
- Use ordered lists and tables to help solve problems systematically. (4Ps5)

Resources

Set of large cards made from photocopiable page 141.

Starter

- Shuffle the set of large cards made from photocopiable page 141. Display one card at a time, asking the learners to call out 'odd' or 'even'. As you do this, sort the odd and even cards into separate piles. Keep the pace brisk.

- Ask: *Which digit did you look at in order to find out whether the number was odd or even?* (The units digit.) *Which units digits belong to even numbers?* (0, 2, 4, 6 and 8.) *Which units digits belong to odd numbers?* (1, 3, 5, 7 and 9.)

Main activities

- Display all the cards from the Starter activity sorted into odd and even.

- Ask the learners to choose three odd numbers from the cards on display and find their total. Ask: *Is the answer even or odd?* Ask them to repeat with another three odd numbers. Ask: *What do you notice?* (The total of three odd numbers is always odd.)

- Draw the following table on the board and ask the learners to copy it. Ask them to investigate the totals of different combinations of odd and even numbers, and record their results in the table.

1st number	2nd number	3rd number	Total
odd	odd	odd	**odd**

- Ask the learners who finish the investigation to write about the patterns they see in their results.

Plenary

- Discuss the results of the investigation.

- Use the results to reinforce the commutative nature of addition; the order in which you add the numbers makes no difference to the total, for example odd + odd + even gives the same result as even + odd + odd and odd + even + odd.

Success criteria

Ask the learners:

- Which of these numbers are even? 156, 7621, 390, 3172, 209, 7723. How can you tell?
- If you add three odd numbers, is the answer odd or even?
- If you add three even numbers, is the answer odd or even?
- *Even + odd + even = even.* Is this statement true? Can you explain how you checked?

Ideas for differentiation

Support: In the investigation, give these learners 'crib sheets' that list the rules for determining odd and even numbers.

Extension: Ask these learners to investigate the odd / even patterns in calculations with the following format: ☐ + ☐ − ☐.

Odd and even cards

50	812	3354
76	498	7260
22	834	6076
80	910	9402
14	546	1268
81	323	8575
27	499	7201
63	135	2697
19	611	4703
55	387	9549

Number sequences 2

Learning objectives

- Describe and continue number sequences, e.g. 7, 4, 1, –2 … identifying the relationship between each number. (4Ps6)
- Recognise and extend number sequences formed by counting in steps of constant size, extending beyond zero when counting back. (4Nn14)

Resources

Counting stick; –40 to 40 number lines; counters (transparent if available); photocopiable page 143.

Starter

- Using a counting stick, start on the fifth division (the middle of the stick), naming your starting point 0. Lead the class in counting on in steps of 2 until you reach the end of the stick (10), then counting back again (10, 8, 6, 4, 2, 0), continuing past 0 until you reach the other end of the stick (–2, –4, –6, –8, –10).
- Repeat the activity, counting in 3s, 4s, 5s and 10s.

Main activities

- Draw a –40 to 40 number line on the board, and give each learner a pre-drawn –40 to 40 number line.
- On the board, write a sequence of four numbers counting on or back in steps of a constant size whose first eight terms are between –40 and 40, for example 15, 11, 7, 3 … Ask the learners to describe the rule (for example 'Count back in fours'). Circle the numbers in the sequence on your number line, asking the learners to mark the same numbers on their number line using counters. Ask them to put counters on the next four numbers in the sequence, for example –1, –5, –9, –13. Repeat for several number sequences.
- Hand out photocopiable page 143, which contains more number sequences to solve and continue. Ask the learners to write the next four terms in each sequence, using the number lines to help them.

Plenary

- On the board, write a number sequence that does not count on or back in steps of a constant size, because it uses multiplication combined with either addition or subtraction, for example 1, 3, 7, 15 …
- Challenge the learners to describe and continue the sequence (for example 'The rule is double and add 1; the next three numbers are 31, 63 and 127').

Success criteria

Ask the learners:

- Can you describe the rule for this number sequence? 6, 9, 12, 15 …
- What are the next three numbers in the sequence?
- Can you describe the rule for this number sequence? 17, 12, 7, 2 …
- What are the next three numbers in the sequence?

Ideas for differentiation

Support: Ask these learners to attempt only the first five questions on photocopiable page 143.

Extension: In the final Main activity, challenge these learners to devise some number sequences that use multiplication as well as addition or subtraction.

Continuing number sequences

1. For each number sequence, work out the rule and write the next four numbers in the sequence.

Count on to create these number sequences.

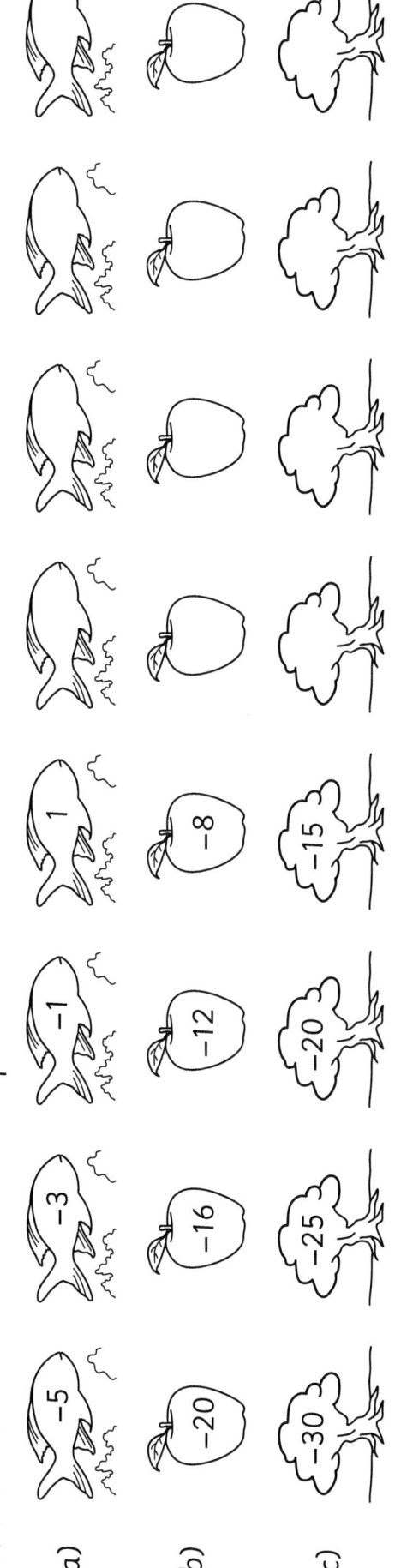

a) –5 –3 –1 1

b) –20 –16 –12 –8

c) –30 –25 –20 –15

Count back to complete these number sequences.

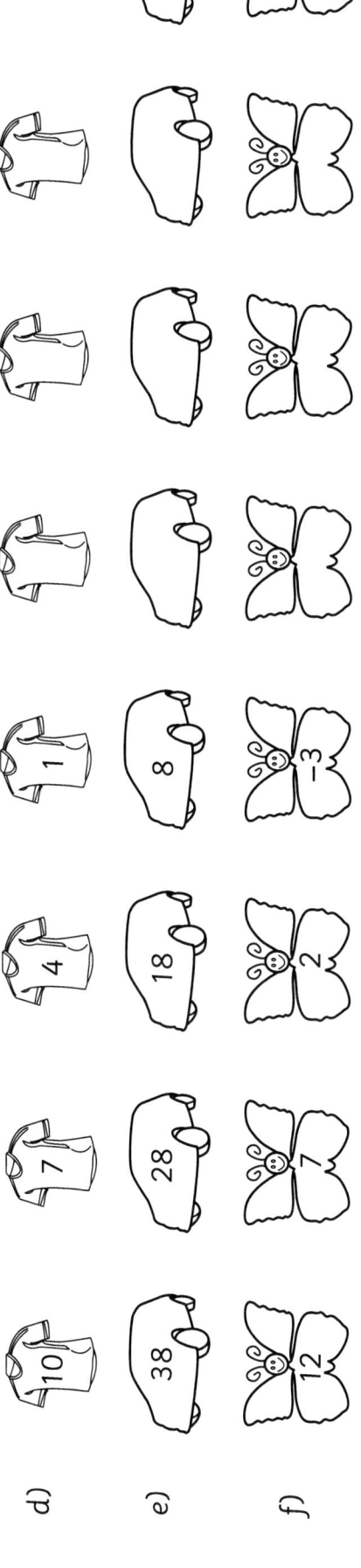

d) 10 7 4 1

e) 38 28 18 8

f) 12 7 2 –3

2. Write some more number sequences. Give them to a friend to solve.

Fractions 1

Learning objectives

● Order and compare two or more fractions with the same denominator (halves, quarters, thirds, fifths, eighths or tenths). (4Nn17)

● Recognise the equivalence between: $\frac{1}{2}$, $\frac{4}{8}$ and $\frac{5}{10}$; $\frac{1}{4}$ and $\frac{2}{8}$; $\frac{1}{5}$ and $\frac{2}{10}$. (4Nn18)

● Use equivalence to help order fractions, e.g. $\frac{7}{10}$ and $\frac{3}{4}$. (4Nn19)

Resources

Cards made from photocopiable page 145; photocopiable page 146.

Starter

• Organise the learners into ability groups of four, six or eight. Ask the learners in each group to pair up. Give each pair a set of cards made from photocopiable page 145.

• Ask pairs within each group to race to match sets of three cards showing a fraction: in diagram form, written in words and written in numerals. The pair who wins the race scores a point.

• Ask the learners to shuffle the cards and play again. The winning pair is the first pair to three points.

Main activities

• Hand out copies of the fraction chart on photocopiable page 146, and display an enlarged version.

• Write a pair of fractions on the board that have the same denominator, for example $\frac{4}{5}$, $\frac{3}{5}$. Ask: *Which fraction is greater? How do you know?* Repeat for other pairs of fractions with the same denominator and extend to ordering sets of three.

• Ask the learners to find one-half on the fraction chart. Ask: *What other fractions are equal to one-half? How do you know?* Introduce the term 'equivalent' to describe equal fractions. Ask: *Can you find a fraction equivalent to one-third / one-quarter / one-fifth?*

• Write a pair of fractions on the board that have different denominators, for example $\frac{3}{4}$ and $\frac{7}{10}$. Ask: *Which fraction is greater? How do you know?* Repeat for similar pairs of fractions, for example $\frac{2}{3}$ and $\frac{3}{5}$; $\frac{3}{5}$ and $\frac{5}{8}$, $\frac{5}{8}$ and $\frac{2}{3}$ and extend to ordering sets of three fractions, for example $\frac{2}{5}$, $\frac{3}{8}$ and $\frac{1}{3}$.

• Ask the learners to devise their own fraction ordering questions and give them to a friend to answer.

Plenary

• Ask selected learners to share one of the fraction ordering questions they devised, and invite the rest of the class to answer it.

• Ask: *How many fractions can you find on the fraction chart that are equivalent to 1? What do you notice about them?*

Success criteria

Ask the learners:

● Can you write these fractions in order from smallest to greatest? $\frac{5}{8}$, $\frac{7}{8}$, $\frac{2}{8}$

● Which fraction is greater: $\frac{6}{8}$ or $\frac{7}{10}$? How do you know?

● Which fractions on your fraction chart are equivalent to one-half?

● Choose three fractions from the fraction chart with different denominators. Can you write them in order, from smallest to greatest?

Ideas for differentiation

Support: Group these learners together. Give them more practice ordering given pairs of fractions before they devise their own questions.

Extension: Challenge these learners, when writing their own fraction ordering questions, to include fractions greater than 1. Give them an extra copy of the fraction chart.

Fraction match cards

	one-third	$\frac{1}{3}$
	one-quarter	$\frac{1}{4}$
	one-tenth	$\frac{1}{10}$
	one-half	$\frac{1}{2}$
	two-fifths	$\frac{2}{5}$
	five-sixths	$\frac{5}{6}$
	three-eighths	$\frac{3}{8}$
	three-quarters	$\frac{3}{4}$
	two-thirds	$\frac{2}{3}$

Fraction chart

1									

$\dfrac{1}{2}$	$\dfrac{1}{2}$

$\dfrac{1}{3}$	$\dfrac{1}{3}$	$\dfrac{1}{3}$

$\dfrac{1}{4}$	$\dfrac{1}{4}$	$\dfrac{1}{4}$	$\dfrac{1}{4}$

$\dfrac{1}{5}$	$\dfrac{1}{5}$	$\dfrac{1}{5}$	$\dfrac{1}{5}$	$\dfrac{1}{5}$

$\dfrac{1}{6}$	$\dfrac{1}{6}$	$\dfrac{1}{6}$	$\dfrac{1}{6}$	$\dfrac{1}{6}$	$\dfrac{1}{6}$

$\dfrac{1}{8}$	$\dfrac{1}{8}$	$\dfrac{1}{8}$	$\dfrac{1}{8}$	$\dfrac{1}{8}$	$\dfrac{1}{8}$	$\dfrac{1}{8}$	$\dfrac{1}{8}$

$\dfrac{1}{10}$	$\dfrac{1}{10}$	$\dfrac{1}{10}$	$\dfrac{1}{10}$	$\dfrac{1}{10}$	$\dfrac{1}{10}$	$\dfrac{1}{10}$	$\dfrac{1}{10}$	$\dfrac{1}{10}$	$\dfrac{1}{10}$

Cambridge Primary: Ready to Go Lessons for Maths Stage 4 © Hodder & Stoughton Ltd 2013

Fractions 2

Learning objectives

● Understand the equivalence between one-place decimals and fractions in tenths. (4Nn20)

● Understand that $\frac{1}{2}$ is equivalent to 0.5 and also to $\frac{5}{10}$. (4Nn21)

● Recognise the equivalence between the decimal fraction and vulgar fraction forms of halves, quarters, tenths and hundredths. (4Nn22)

Resources

Calculators; cards made from photocopiable page 148.

Starter

● Write a unitary fraction on the board; a fraction whose numerator is 1, for example $\frac{1}{6}$. Explain that the line in a fraction means 'divided by'. On the board, write '$\frac{1}{6} = 1 \div 6$'.

● Call out unitary fractions one at a time. For each fraction, ask the learners to write the fraction in numerals and write the division it represents.

Main activities

● Organise the learners into pairs, giving each pair a calculator and a set of cards made from photocopiable page 148.
  ● Ask the learners to place the fraction cards face down in a single pile and spread the decimal cards around, face up.
  ● Ask them to take it in turns to turn over the top card on the pile, for example $\frac{3}{10}$, and use the calculator to divide the numerator by the denominator, for example $3 \div 10$.
  ● Ask them to find the decimal card that matches the calculator display, for example 0.3, and pair the decimal card with the fraction card, for example 0.3 and $\frac{3}{10}$.

● After pairs have finished matching the fraction and decimal cards, ask them to describe and explain any patterns (for example the numerator in a tenths fraction is always the same as the digit after the decimal point in the decimal, because they both represent the same fraction – tenths).

● Ask pairs to use the cards to play either a game of snap or a game of memory.

Plenary

● Write 0.5 on the board. Ask: *How would you write this as a fraction with a denominator of 10? What other fractions are equivalent to 0.5? How do you know?*

● Ask the learners to find fractions equivalent to 0.25 and 0.75, and describe their methods.

Success criteria

Ask the learners:

● What is 0.7 written as a fraction?

● What is $\frac{4}{10}$ written as a decimal?

● Can you write two fractions that are equivalent to 0.5?

● What is one-quarter written as a decimal? How do you know?

Ideas for differentiation

Support: Support these learners when they are playing their game of snap or memory.

Extension: Ask these learners to investigate the relationship between fractions with a denominator of 100 and decimals, for example $\frac{32}{100} = 0.32$.

Tenths in fractions and decimals

$\frac{1}{10}$	$\frac{2}{10}$	$\frac{3}{10}$	$\frac{4}{10}$
$\frac{5}{10}$	$\frac{6}{10}$	$\frac{7}{10}$	$\frac{8}{10}$
$\frac{9}{10}$	$\frac{10}{10}$	0.1	0.2
0.3	0.4	0.5	0.6
0.7	0.8	0.9	1

 Cambridge Primary: Ready to Go Lessons for Maths Stage 4 © Hodder & Stoughton Ltd 2013

Fractions 3

Learning objectives

● Identify simple fractions with a total of 1, e.g. $\frac{1}{4} + \boxed{} = 1$. (4Nc3)

● Recognise mixed numbers, e.g. $5\frac{3}{4}$, and order these on a number line. (4Nn23)

Resources

Photocopiable page 146; large blank number line with 20 divisions; cards made from photocopiable page 150; squared paper.

Starter

• Display an enlarged version of photocopiable page 146 (the fraction chart) and hand out copies to the learners.

• On the board, write a missing number addition in which the total is 1, for example $\frac{2}{3} + \boxed{} = 1$. Ask the learners to work out the missing number, for example $\frac{1}{3}$, and explain their method. Ask: *Is there another fraction you could write in the box?* (Yes, for example $\frac{2}{6}$.) Discuss why.

• Repeat for other similar number sentences, for example $\boxed{} + \frac{2}{4} = 1$; $\frac{3}{5} + \boxed{} = 1$; or $\boxed{} + \frac{1}{2} = 1$.

Main activities

• Display the blank number line with 20 divisions. Label the left end 0, the right end 5 and every fourth division with a whole number. Ask the learners to help you write in the half numbers. Next, point to the division halfway between 0 and $\frac{1}{2}$ and ask: *What's this number?* Establish that the smallest divisions are quarters. Label $\frac{1}{4}$ and $\frac{3}{4}$ on the number line. Ask volunteers to label the rest of the quarter divisions.

• Display six cards made from photocopiable page 150. Model finding each number on the line, and ordering the cards from smallest to largest.

• Ask the learners to make their own fraction number line by copying the number line from the board onto squared paper. Organise the learners into pairs and give each pair a set of cards made from photocopiable page 150. Ask them to turn over six cards, arrange the cards in order from smallest to largest, and then write down this ordered list of numbers.

Plenary

• Challenge the learners to sketch a number line from 0 to 4 marked off in fifths.

• On the board, write a set of six numbers up to 4 that are multiples of one-fifth, asking the learners to order them from smallest to greatest, for example $1\frac{2}{5}$, 4, $\frac{3}{5}$, $2\frac{1}{5}$, $1\frac{4}{5}$, $3\frac{3}{5}$.

Success criteria

Ask the learners:

● What's the missing number in this number sentence? $\frac{1}{3} + \boxed{} = 1$

● How much do you need to add to $\frac{3}{4}$ to make 1? How did you work out the answer?

● Can you draw a number line from 0 to 5 marked in quarters?

● Draw four cards. Read the number on each card aloud. Put the numbers in order from smallest to largest.

Ideas for differentiation

Support: Support these learners in the final Main activity by pairing them with a more confident partner.

Extension: In the final Main activity, challenge these learners to order the numbers without referring to the number line, using it only as a checking tool.

Fraction cards

$\frac{1}{4}$	$\frac{1}{2}$	$\frac{3}{4}$	1
$1\frac{1}{4}$	$1\frac{1}{2}$	$1\frac{3}{4}$	2
$2\frac{1}{4}$	$2\frac{1}{2}$	$2\frac{3}{4}$	3
$3\frac{1}{4}$	$3\frac{1}{2}$	$3\frac{3}{4}$	4
$4\frac{1}{4}$	$4\frac{1}{2}$	$4\frac{3}{4}$	5

Cambridge Primary: Ready to Go Lessons for Maths Stage 4 © Hodder & Stoughton Ltd 2013

Fractions 4

Learning objectives

● Relate finding fractions to division. (4Nn24)
● Find halves, quarters, thirds, fifths, eighths and tenths of shapes and numbers. (4Nn25)
● Explain methods and reasoning orally and in writing. (4Ps9)

Resources

Beanbag or soft ball; four identical 8 × 3 rectangles drawn on a large square grid; two identical 5 × 10 rectangles drawn on a large square grid; photocopiable page 152.

Starter

- Throw a beanbag or soft ball to one learner at a time, asking the learner a times tables division question, for example: *What is 18 ÷ 3?* Ask division questions from different times tables to match the abilities of the learners.

- The learner must throw the beanbag or ball back to you as they answer the question, and then you throw it to another learner, asking them a different question. Keep the pace brisk.

Main activities

- Display one of the 8 × 3 rectangles. Ask: *What number does this rectangle represent?* (Twenty-four.) Ask a volunteer to shade one-half of the rectangle. Ask: *What's one-half of twenty-four?* (Twelve.) Remind the learners that finding one-half is equivalent to dividing by two. Use the other three copies of the rectangle to find one-third (÷ 3), one-quarter (÷ 4) and one-eighth (÷ 8).

- Display one of the 5 × 10 rectangles. Use it in a similar way to link finding one-fifth with dividing by 5, and finding one-tenth with dividing by 10.

- Model finding unit fractions of numbers using division facts only, for example $\frac{1}{3}$ of 15 = 5 because 15 ÷ 3 = 5. Give the learners similar questions to answer, finding $\frac{1}{2}$, $\frac{1}{3}$, $\frac{1}{4}$, $\frac{1}{5}$, $\frac{1}{8}$ and $\frac{1}{10}$ of numbers (whole-number answers only).

- Hand out photocopiable page 152 and read through the word problems together with the learners. Ask them to solve the problems, working individually or in pairs.

Plenary

- Go through the answers to the problems on photocopiable page 152.

- Ask the learners to explain how they worked out each answer.

Success criteria

Ask the learners:

● Find a half of this shape.
● Finding a sixth of a number is the same as dividing the number by what?
● What number do I need to divide 80 by to find one eighth of it?
● What is $\frac{1}{4}$ of 28?

Ideas for differentiation

Support: Ask these learners to attempt only the first five problems on photocopiable page 152, as these cover the more familiar times tables.

Extension: Challenge these learners to write their own word problems that involve finding fractions of numbers.

Name: _____

Finding fractions of numbers

1. There are 70 children in Year 4.
    One-tenth of them were born in June.
    How many children were born in June?

    _____

2. Abdulah has 18 figs, but only half of them are ripe.
    How many figs are ripe?

    _____

3. There are 30 children in Class 4.
    One-fifth of them go home for lunch.
    How many children go home for lunch?

    _____

4. A bucket holds 15 litres of water when it is full.
    It is one-third full.
    How much water is in the bucket?

    _____

5. Charisse baked 16 cupcakes.
    Her friends ate one-quarter of them.
    How many cupcakes did Charisse's friends eat?

    _____

6. A book has 64 pages.
    Leroy reads one-eighth of the book.
    How many pages does he read?

    _____

7. Three sisters weigh a total of 42 kg.
    The youngest sister, a baby, weighs $\frac{1}{6}$ of the total.
    How much does the baby sister weigh?

    _____

8. Nadia is saving up to buy a music player that costs $45.
    So far she has saved one-ninth of the cost.
    How much money has she saved so far?

    _____

Cambridge Primary: Ready to Go Lessons for Maths Stage 4 © Hodder & Stoughton Ltd 2013

Addition and subtraction facts

Learning objectives

- Derive quickly pairs of two-digit numbers with a total of 100, e.g. 72 + ☐ = 100. (4Nc1)
- Derive quickly pairs of multiples of 50 with a total of 1000, e.g. 850 + ☐ = 1000. (4Nc2)

Resources

Counting stick; 0 to 9 digit cards; timers; calculators; cards made from photocopiable page 154.

Starter

- On a counting stick, count from 0 to 100 in 5s, using each half division to represent 5.

- Point to multiples of 5 on the counting stick in ascending order (5, 10, 15, 20 ...), asking each time: *How many more to make 100?* (95, 90, 85, 80 ...)

- Point to multiples of 5 on the counting stick in a random order, asking: *What number is this?* (For example forty-five.) *How many more to make 100?* (For example fifty-five.)

Main activities

- Use knowledge of pairs of multiples of 5 totalling 100 to derive pairs of any two-digit numbers with a total of 100, for example use knowledge that 15 + 85 = 100 to work out 14 + ☐ = 100. Check by partitioning and recombining.

- Organise the learners into groups. Give each group a set of 0 to 9 digit cards, a timer and a calculator. Ask the learners to use the cards to generate a two-digit number, and set the timer (for example for 20 seconds). The learners must write down the generated number's complement to 100. Every correct answer scores a point. The learners can use the calculator to check.

- Use knowledge of pairs of multiples of 5 totalling 100 to derive pairs of multiples of 50 that total 1000, for example use knowledge that 25 + 75 = 100 to work out 250 + ☐ = 1000. Check by partitioning and recombining.

- Give each group a set of cards made from photocopiable page 154. Ask the learners to play the same game as before, turning over a card from photocopiable page 154 to generate each number, and writing down the complement to 1000.

Plenary

- Use a set of 0 to 9 digit cards to give the learners quick-fire questions asking them to derive pairs of two-digit numbers with a total of 100.

- Use a set of cards from photocopiable page 154 to do the same for pairs of multiples of 50 with a total of 1000.

Success criteria

Ask the learners:

- What is the missing number in this number sentence: 350 + ☐ = 1000?
- How did you work out your answer?
- How many more do you need to add to 46 to make 100?
- How could you check that your answer is correct?

Ideas for differentiation

Support: Give these learners a list of pairs of multiples of 5 that total 100.

Extension: Group these learners together and ask them to use a more challenging time limit in the games (for example 10 seconds).

Multiples of 50 cards

50	100	150	200
250	300	350	400
450	500	500	550
600	650	700	750
800	850	900	950

Cambridge Primary: Ready to Go Lessons for Maths Stage 4 © Hodder & Stoughton Ltd 2013

Addition strategies 4

Learning objectives

- Add three two-digit multiples of 10, e.g. 40 + 70 + 50. (4Nc7)
- Add and subtract near multiples of 10 or 100 to or from three-digit numbers, e.g. 367 – 198 or 278 + 49. (4Nc8)
- Check the results of adding numbers by adding them in a different order or by subtracting one number from the total. (4Pt3)

Resources

Two large sets of multiples of 10 cards (10–90); 0 to 9 digit cards; cards made from photocopiable page 156.

Starter

- Shuffle together the multiples of 10 cards. Draw three cards and display them (for example 30, 30, 70). Ask the learners to write the total.
- Discuss strategies used. These might include: looking for pairs that make 100 (for example 30 + 70); using doubles knowledge (for example double 30 is 60) or using single-digit addition facts and multiplying by 10 (for example 3 + 3 + 7 = 13 so 30 + 30 + 70 = 130).

Main activities

- On the board, write an addition containing a three- or four-digit number, and a multiple of 10, 100 or 1000, for example 238 + 90 or 184 + 3000. Ask the learners to find the total and describe their method. Ask: *How could you check your answer?* (For example add the numbers in a different order or subtract one number from the total.) Repeat.
- On the board, write an addition containing a three-digit number and a near multiple of 10 or 100, for example 743 + 39 or 432 + 501. Discuss strategies, including compensation from multiples of 10 and 100. Repeat.

- Organise the pairs into groups. Give each group a set of digit cards and a set of cards made from photocopiable page 156. Ask the learners to generate a three-digit number using the digit cards, turn over a card from photocopiable page 156, and add the numbers together. They should then check each other's answers. Every learner with the correct answer scores a point.

Plenary

- On the board, write some additions containing a three-digit number and a near multiple of 10 or 100. Include answers, some of which should be incorrect.
- Challenge the learners to identify the correct answers. Ask them to explain how they checked the answers.

Success criteria

Ask the learners:

- What's 20 + 60 + 80?
- What's 832 + 4000?
- What's 466 + 197? How did you work out the answer?
- How could you check your answer?

Ideas for differentiation

Support: Give these learners sets of place value cards (thousands, hundreds, tens and units) to help them with their calculations.

Extension: Ask these learners to use the 0 to 9 digit cards to generate some four-digit numbers as well as three-digit numbers.

Near multiples of 10 and 100

51	92	81
79	28	49
62	38	101
702	903	201
502	899	398
697	599	498

Cambridge Primary: Ready to Go Lessons for Maths Stage 4 © Hodder & Stoughton Ltd 2013

Subtraction strategies 3

Learning objectives

- Find a difference between near multiples of 100, e.g. 304 – 296. (4Nc11)
- Subtract a small number crossing 100, e.g. 304 – 8. (4Nc12)
- Choose strategies to find answers to addition or subtraction problems; explain and show working. (4Ps3)
- Check subtraction by adding the answer to the smaller number in the original calculation. (4Pt4)

Resources

Photocopiable page 158.

Starter

- On the board, write a calculation involving subtracting a single-digit number from a three-digit number, crossing the hundreds boundary, for example 403 – 9 = 394. Ask the learners to check the answer, making a thumbs-up sign if it is correct, and making a thumbs-down sign if it is incorrect.
- Ask the learners to describe how they checked the answer (for example by adding the answer to the smaller number in the original calculation).
- Repeat several times. Include some calculations in which the answer is incorrect.

Main activities

- Hand out photocopiable page 158 and read the first word problem. Ask: *What calculation do you need to do?* (402 – 7.) Ask the learners to find the answer and explain their method. Model strategies, including counting back on a number line. Ask: *How could you check your answer?* Model adding the answer to the smaller number in the original calculation.

- Read the second word problem on photocopiable page 158. Establish that the calculation required is 706 – 699. Ask: *Would starting at 706 and counting back 699 be a good strategy?* (No.) *Why not?* (Because it would take ages and you could easily lose count.) Ask the learners to suggest an alternative strategy (for example locating 706 and 699 on a number line and finding the difference). Ask the learners to check their answer by adding it to the smaller number in the original calculation.
- Ask the learners to solve the remainder of the problems on photocopiable page 158, recording their working.
- Ask the learners who finish to check each other's answers.

Plenary

- For each of the remaining problems on photocopiable page 158, ask the learners to describe the calculation they did, the method they used and the answer they got.
- Ask the rest of the class to confirm whether a given answer is correct, explaining how they know.

Success criteria

Ask the learners:

- What's 205 – 7? How did you work it out?
- What calculation could you do to check your answer?
- What's 304 – 296? How did you work it out?
- What calculation could you do to check your answer?

Ideas for differentiation

Support: Group these learners together and work through a few more problems on photocopiable page 158 with them.

Extension: Ask these learners to write their own word problems featuring similar subtractions.

Name: _____

Subtraction problems

1. The distance between two towns is 402 kilometres
   via Route A. The distance via Route B is 7 kilometres less.
   How long is Route B?

   _____

2. A jug contans 706 ml of fruit juice.
   Two children have a drink, pouring out
   a total of 699 ml of juice from the jug.
   How much fruit juice is left in the jug?

   _____

3. A factory produces 503 mobile phones in a single day.
   Nine of them are faulty, and do not pass quality control.
   How many mobile phones pass quality control?

   _____

4. There are 308 tickets for a concert.
   295 tickets are sold on the first day.
   How many tickets are left?

   _____

5. In a game you score 604.
   Your friend scores 593.
   How much greater is your score than your friend's?

   _____

6. You have $101 in your savings account.
   You take out $93.
   How much money do you have left in your savings account?

   _____

7. One year scientists count 205 penguins in a particular colony.
   The next year they count eight fewer penguins.
   How many penguins do scientists count the second year?

   _____

8. 403 guests are invited to a wedding.
   Six of them are unable to attend.
   How many guests are there at the wedding?

   _____

Cambridge Primary: Ready to Go Lessons for Maths Stage 4 © Hodder & Stoughton Ltd 2013

Doubling and halving 2

Learning objectives

- Derive quickly doubles of all whole numbers to 50, doubles of multiples of 10 to 500, doubles of multiples of 100 to 5000, and corresponding halves. (4Nc16)
- Explore and solve number problems and puzzles. (4Ps4)

Resources

Coloured sticky notes; photocopiable page 160.

Starter

- Organise the learners into two or more teams and sit each team in a line. Put a pencil and some sticky notes at the front of each line, using a different colour of note for each team.
- On the board, write either: a whole number below 50, asking the learners to double it; or write an even number between 50 and 100, asking the learners to halve it. The learners at the front of each line must write their answer on a sticky note and race to hand it to you, then go to the back of the line.
- Keep the first correct sticky note you receive each round and discard the others. At the end of the game, the winners are the team with the greatest number of kept sticky notes.

Main activities

- Write on the board: Double 37 = ☐
- Ask the learners to find the missing number (74), and explain how they worked it out.
- Underneath, write: Double 370 = ☐
  Double 3700 = ☐
- Ask the learners to fill in the missing numbers, explaining their methods. Explore other similar families of doubling facts.

- Repeat the activity for corresponding families of halving facts (for example half of 58 / half of 580 / half of 5800).
- Hand out photocopiable page 160. Ask the learners to place one number in each circle, so that the total of the numbers on the long axis is double the total of the numbers on the short axis.

Plenary

- Discuss solutions to the puzzle on photocopiable page 160.
- Ask the more-able learners who have devised their own doubling or halving puzzles to present them to the class.

Success criteria

Ask the learners:

- What's double 49? How did you work it out?
- What's half of 760? How did you work it out?
- Which doubling fact can help you work out the answer to double 3200?
- How could you check that half of 76 equals 38?

Ideas for differentiation

Support: Give these learners a modified version of the puzzle using the numbers 10, 20, 30, 40 and 50. (Correct solutions have 30 in the central circle, with axes totalling 60 and 120.)

Extension: Ask these learners to devise their own doubling or halving puzzles involving multiples of 10 or multiples of 100.

Name: _____

Doubling puzzle

1. Write one number in each circle in the diagram below so that the total of the numbers going across is double the total of the numbers going down.

    Use these numbers:

    | 120   160   200   340   420 |

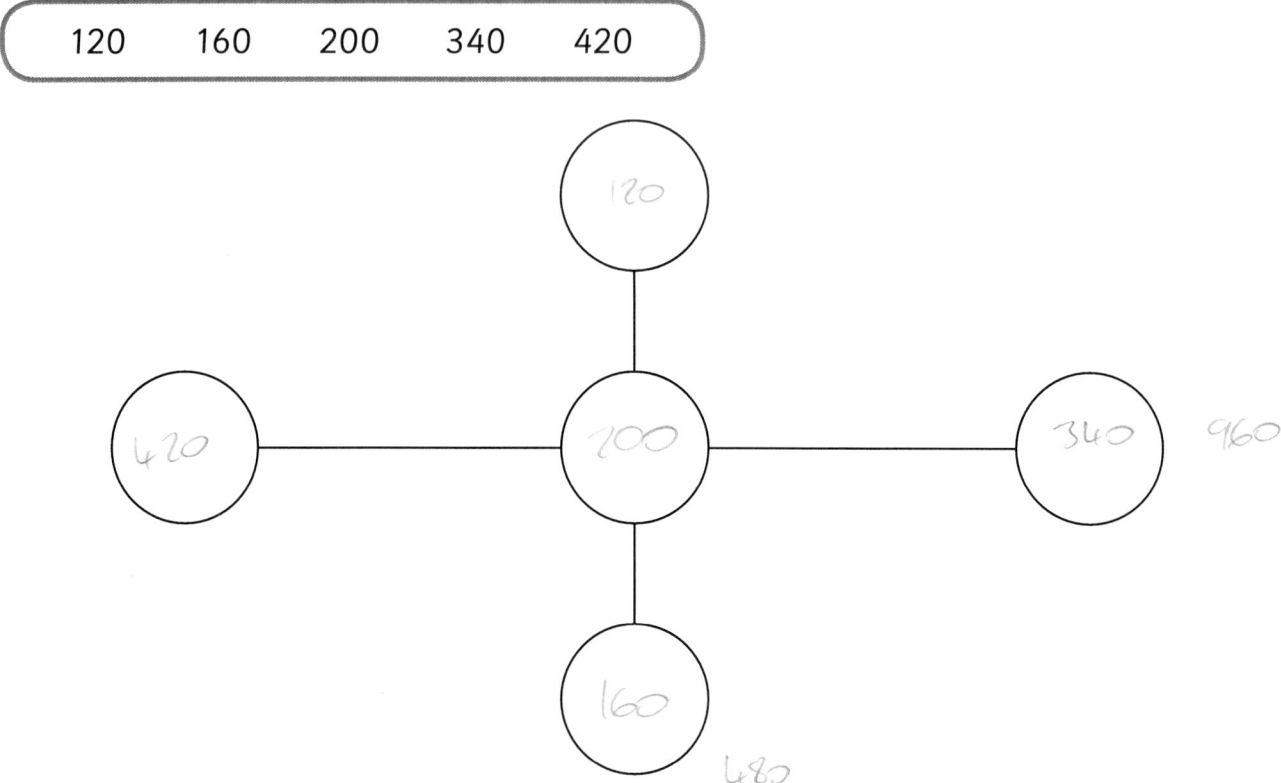

120

420   200   340   960

160

480

2. Can you create and solve another puzzle by using your own numbers?

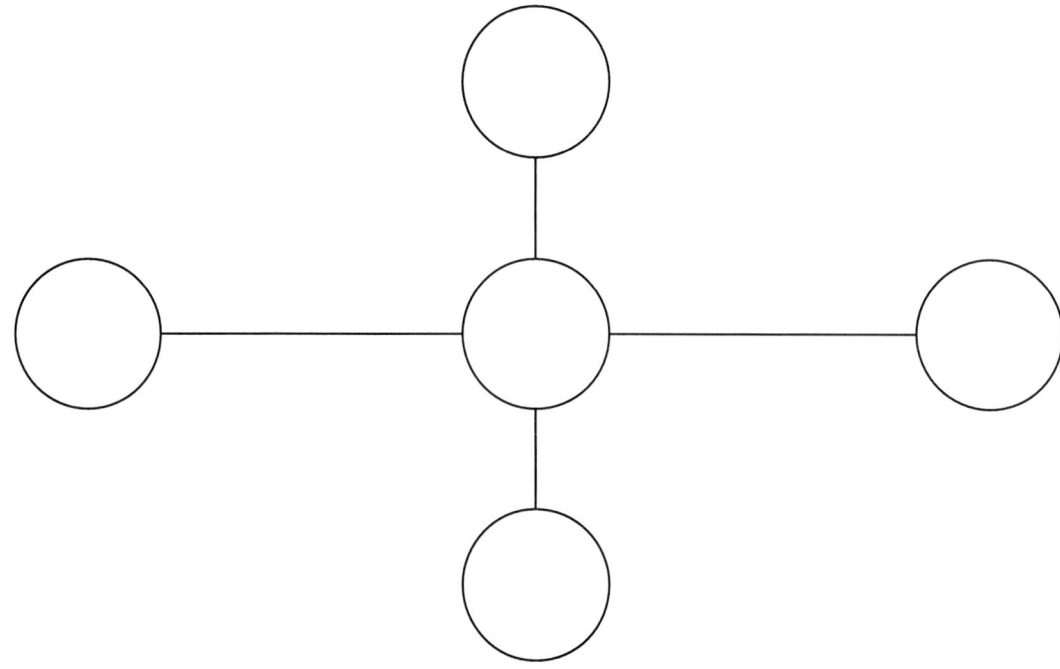

Cambridge Primary: Ready to Go Lessons for Maths Stage 4 © Hodder & Stoughton Ltd 2013

Multiplication and division strategies

Learning objectives

- Explain reasons for a choice of strategy when multiplying or dividing. (4Ps2)
- Check multiplication using a different technique, e.g. check 6 × 8 = 48 by doing 6 × 4 and doubling. (4Pt5)
- Check the result of a division using multiplication, e.g. multiply 4 by 12 to check 48 ÷ 4. (4Pt6)
- Decide whether to round up or down after division to give an answer to a problem. (4Nc24)
- Begin to understand simple ideas of ratio and proportion, e.g. a picture is one fifth the size of the real dog. It is 25 cm long in the picture, so it is 5 × 25 cm in real life. (4Nc26)

Resources

Photocopiable page 162; multiplication grid made from photocopiable page 93.

Starter

- Give the learners multiplications and divisions (no remainders), asking them to write down the answers. Ask them to explain their methods and why they chose them.
- Ask them to check multiplications using a different technique (for example check that 12 × 6 = 72 by multiplying 6 by 6 and then doubling). Ask them to check divisions using multiplication (for example check that 77 ÷ 7 = 11 by multiplying 11 by 7).

Main activities

- Ask the learners the following problem: *The proportion of red balloons in a pack is 1 in every 5. Each pack has eight red balloons. How many balloons of all colours are there in total in a pack?* Explain that when you look at the proportion of an amount, it is the same as finding the fraction of the whole amount: $\frac{1}{5}$ of the balloons are red. Draw this table to help solve the problem:

Red balloons	Total number of balloons
1	5
2	10
4	20
8	?

- Discuss the results, using cubes or counters to help with the understanding and then repeat the activity for other proportions.
- Hand out photocopiable page 162. Read through the questions together, asking the learners to identify which problems can be solved with multiplication and which with division.

Plenary

- Go through the answers to the remainder of the problems on photocopiable page 162.
- Ask the learners to identify which questions involved a division calculation in which there was a remainder, but for which the answer had to be rounded to a whole number (questions 6 and 7). Ask them to say whether they rounded each answer up or down, and explain why.

Success criteria

Ask the learners:

- What is 68 ÷ 5?
- How did you work out the answer?
- How could you check your answer?
- If the question the calculation came from was: 'How many miles do you need to cycle each day to cycle at least 68 miles between Monday and Friday?', what answer would you give? Why?

Ideas for differentiation

Support: Give these learners the multiplication grid made from photocopiable page 93 to use as a reference.

Extension: Ask these learners to work independently to solve the problems.

Name: _____

Multiplication and division problems

1. The proportion of caramels in a box of chocolates is 1 in every 7.

    Each box of chocolates has three caramels.

    How many chocolates are there in a box?

    _____

2. In a diagram, a bicycle is drawn at one-tenth of its actual size.

    In the diagram the bicycle is 18 cm long.

    What is the length of the real bicycle?

    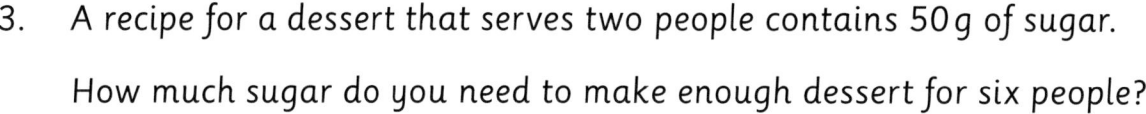

    _____

3. A recipe for a dessert that serves two people contains 50 g of sugar.

    How much sugar do you need to make enough dessert for six people?

    _____

4. Leila's mum is 34 years old. Leila's great-great-grandmother is three

    times Leila's mum's age. How old is Leila's great-great-grandmother?

    _____

5. In a particular house, the area of the bathroom is one-quarter of

    the area of the living room. The living room has an area of 36 m².

    What is the area of the bathroom?

    _____

6. Cans of cherry cola come in packs of six.

    How many full packs can be made from 57 cans?

    _____

7. Pete is making a paper chain. Each chain link is 9 cm wide.

    What is the fewest number of chain links he needs to join together

    to make a chain at least 75 cm long?

    _____

Cambridge Primary: Ready to Go Lessons for Maths Stage 4 © Hodder & Stoughton Ltd 2013

Unit assessment

Questions to ask

- What is 503 – 8? How did you work out the answer?
- What number do you need to add to 63 to make 100? How did you work out the answer?
- Can you write three fractions that are equivalent to one-half?

- How would you write 245 cm in metres?
- What's one-quarter of 44? How did you work it out?
- How would you write $8.50 in cents?

Summative assessment activities

Observe the learners while they take part in these activities. You will quickly be able to identify those who appear to be confident and those who may need additional support.

Making 1

This game assesses the learners' ability to understand the equivalence between one-place decimals and fractions in tenths, and identify simple fractions with a total of 1.

You will need:

Cards made from photocopiable page 148.

What to do

- Organise the learners into groups of three.
- Put to one side the '$\frac{10}{10}$' and '1' cards from a set of cards made from photocopiable page 148, and shuffle the remaining cards together. Deal out six cards to each player. Decide who will be the start player. The start player must choose a card from their hand and place it face up on the table (for example $\frac{4}{10}$).
- The player to the start player's left must play a card that makes 1 when added to the face-up card (for example 0.6 or $\frac{6}{10}$). If the player cannot play a card, play passes to the next player. Whichever player completes the pair of cards plays the first card in the next pair. The winner is the first player with no cards left.

Making 1000 memory game

This game assesses the learners' ability to derive quickly pairs of multiples of 50 with a total of 1000.

You will need:

Cards made from photocopiable page 154.

What to do

- Organise the learners into groups of two to four.
- Give each group a set of cards made from photocopiable page 154. Ask them to shuffle the cards and place them face down in a 4 × 5 rectangular grid.
- Tell them to take it in turns to turn over two cards. If the cards they turn over total 1000 they keep the cards.
- The winner is the player with the most cards when there are no more cards left in the grid.

Written assessment

Distribute photocopiable page 164. Ask the learners to read the questions and write the answers. They should work independently.

Solve the problems

1. Tim has an arm span of 128 cm. What is this in metres?

    _____

    a) Tim's friend William has an arm span of 1.3 m.

        What is this in centimetres? _____

    b) What is the friends' total arm span in centimetres? _____

    c) What is their total arm span in metres? _____

2. The temperature in a refrigerator is 6°C. The temperature in the refrigerator's freezer compartment is 19°C lower. What is the temperature in the freezer compartment?

    _____

3. Write the next three numbers in each number sequence.

    a) 2, 5, 8, _____, _____, _____

    b) 5, 3, 1, _____, _____, _____

    c) 158, 168, 178, _____, _____, _____

4. Order each set of fractions from smallest to greatest:

    a) $\frac{2}{10}, \frac{9}{10}, \frac{7}{10}, \frac{5}{10}, \frac{1}{10}, \frac{4}{10}$ _____

    b) $\frac{5}{8}, \frac{3}{8}, \frac{7}{8}, \frac{2}{8}, \frac{6}{8}, \frac{4}{8}$ _____

    c) $\frac{1}{2}, \frac{1}{6}, \frac{1}{3}, \frac{1}{5}, \frac{1}{4}$ _____

5. Draw lines to match each fraction in the top row to the matching decimal fraction in the bottom row.

$\frac{1}{2}$	$\frac{3}{4}$	$\frac{1}{4}$	$\frac{1}{10}$	$\frac{1}{5}$
0.2	0.75	0.1	0.2	0.25

6. Raphael and his brother have collected a total of 457 stickers. Raphael has collected 298 stickers. How many stickers has his brother collected?

    _____

Cambridge Primary: Ready to Go Lessons for Maths Stage 4 © Hodder & Stoughton Ltd 2013

Unit 3B: Measure and problem solving

Time 7

Learning objectives

● Read and tell the time to the nearest minute on 12-hour digital and analogue clocks. (4Mt1)

● Use am, pm and 12-hour digital clock notation. (4Mt2)

Resources

Analogue clock faces with movable hands; cards made from photocopiable pages 47 and 48; photocopiable page 166.

Starter

- Hand out the clock faces, one per learner.
- On the board, write a digital time that is past the half hour, for example 2:43. Ask the learners to make the time on their analogue clock faces.
- Ask the learners to say the time in two different ways: as a 'past' time (for example forty-three minutes past two) and as a 'to' time (for example seventeen minutes to three).
- Repeat for other similar times, for example 8:57, 11:38, 4:41, 1:36, 9:52.

Main activities

- Organise the learners into pairs. Give each pair photocopiable page 166. Explain that the circle represents the 24 hours in a day.
- On the board, write a non-ordered list of events, for example:
  a) break time
  b) midnight
  c) the time you left home this morning
  d) the time school finishes
  e) midday
  f) the time this lesson started.
  Ask the learners to indicate each time on the circle. Discuss responses. Ask the learners to add their own events.

- Organise the learners into groups of six and give each group a set of cards made from photocopiable pages 47 and 48.
  - Ask them to shuffle the cards and deal six to each player.
  - The starting player should place a card face up.
  - Play passes to the left.
  - Any player with a matching time can place it face up next to the first card.
  - When three matching cards have been played, they can be turned face down and the next player to the left should choose a card to start a new set of three.
  - The winner is the first player to get rid of all their cards.

Plenary

- Write a time on the board to the nearest minute in digital notation, including a.m. or p.m.
- Ask the learners to make the time on their analogue clock face, indicate the time on their copy of photocopiable page 166, and suggest an event that might happen at that time.

Success criteria

Ask the learners:

● Can you make 12:39 on a clock face?

● Can you say the time you just made in two different ways?

● When might you eat your lunch: 1.00p.m. or 1.00a.m.?

● Can you show me what time you get home from school on photocopiable page 166?

Ideas for differentiation

Support: In the final Main activity, group these learners together, and help them to play the game.

Extension: Ask these learners to find out about the 24-hour clock, and modify the diagram on photocopiable page 166 to include 24-hour times.

Name: _____

Time: a.m. and p.m.

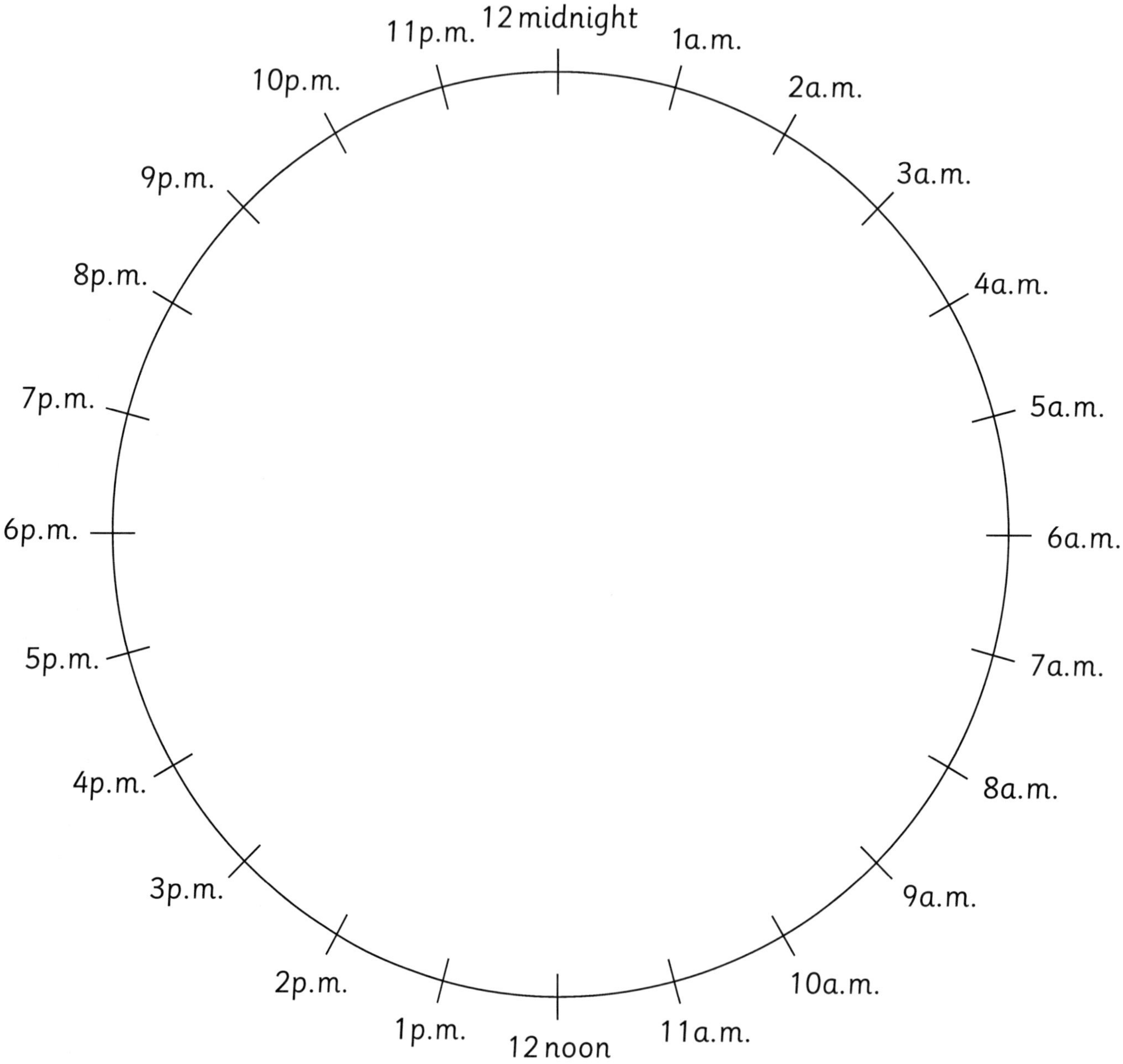

11 p.m. 12 midnight 1 a.m.

10 p.m.   2 a.m.

9 p.m.   3 a.m.

8 p.m.   4 a.m.

7 p.m.   5 a.m.

6 p.m.   6 a.m.

5 p.m.   7 a.m.

4 p.m.   8 a.m.

3 p.m.   9 a.m.

2 p.m.   10 a.m.

1 p.m. 12 noon 11 a.m.

Cambridge Primary: Ready to Go Lessons for Maths Stage 4 © Hodder & Stoughton Ltd 2013

Time 8

Learning objectives

- Read simple timetables and use a calendar. (4Mt3)
- Understand everyday systems of measurement in length, weight and capacity and time and use these to solve simple problems as appropriate. (4Pt2)
- Explain methods and reasoning orally and in writing; make hypotheses and test them out. (4Ps9)

Resources

Calendar for the current year with significant events marked on it; photocopiable page 168; analogue clock faces with movable hands.

Starter

- Organise the learners into pairs, giving each pair a copy of a calendar for the current year, with significant national, local, school and class events marked on it.
- Ask the learners questions, the answers to which can be found by reading and interpreting the calendar, for example you might ask the learners:
  - the length of time between two marked events
  - the length of time since a past event
  - the length of time until a future event
  - to give the date of an unmarked event by telling them how long before or after a particular marked event it is.

Main activities

- Hand out photocopiable page 168 and analogue clock faces with movable hands.
- Read through the first problem and ask the learners to work with a partner to solve it. Ask them to give the answer, and explain how they worked it out (for example by making 7:40 on the clock, then moving the hands on 30 minutes to give 8:10, and then finding the films with showings between 7:40 and 8:10). Repeat the procedure with the second problem.

- Ask the learners to solve the rest of the problems on photocopiable page 168, either continuing to work with a partner, or working independently. When they have finished, challenge them to write their own problems based on the cinema timetable and give them to a friend to solve.

Plenary

- Go through the answers to the final three problems on photocopiable page 168, asking the learners to explain how they solved them.
- Ask selected learners to share the word problems they have written, and ask the rest of the class to solve them.

Success criteria

Ask the learners:

- (Using the calendar:) How long is it since / until 15 August?
- (Using the calendar:) What date is it four weeks and six days after 13 February?
- (Using the cinema timetable on photocopiable page 168:) Which film has the earliest showing of the day?
- (Using the cinema timetable on photocopiable page 168:) Journey to Mars is 1 hour 39 minutes long. What time does the 8:30p.m. showing finish?

Ideas for differentiation

Support: Allow these learners to work with a partner, and ask them to write only a couple of word problems of their own.

Extension: Challenge these learners to work independently, and to write half a dozen word problems of their own.

Name: _____

Cinema timetable problems

The timetable below shows the films showing during one week
at the Kohinoor Cinema.

Film	Times
Journey to Mars	12:50p.m., 3:15p.m., 5:45p.m., 8:30p.m.
Dinosaur Dreams	12:30p.m., 2:40p.m., 4:50p.m., 7:50p.m.
Judo Jaguar	1:05p.m., 3:00p.m., 5:10p.m., 7:20p.m.
The Detective Club	1:15p.m., 2:50p.m., 4:45p.m., 8:05p.m.

1. Dan walks past the cinema at 7:40p.m. He decides to go in and
    watch a film. Which films start within the next half hour?

    _____

2. Aamina turns up a quarter of an hour late for the 3:00p.m. showing of
    Judo Jaguar. How long will she have to wait until the next showing?

    _____

3. Dinosaur Dreams is 1 hour 34 minutes long. What time
    will the 4:50p.m. showing of Dinosaur Dreams finish?

    _____

4. The Detective Club is 1 hour 27 minutes long.
    Mansoor wants to watch The Detective Club,
    but must be out of the cinema before 4:05p.m.
    Can he go to the 2:50p.m. showing?

    _____

5. Journey to Mars is so popular that the manager of the cinema
    adds an extra showing on a different screen.
    The extra showing starts 55 minutes before the final showing.
    What time does the extra showing start?

    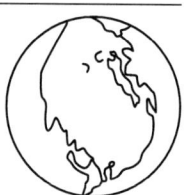

    _____

6. Write your own problems based on the cinema timetable.
    Give them to a friend to solve.

Cambridge Primary: Ready to Go Lessons for Maths Stage 4 © Hodder & Stoughton Ltd 2013

Time 9

Learning objectives

- Choose units of time to measure time intervals. (4Mt4)
- Estimate and approximate when calculating, and check working. (4Pt8)

Resources

Photocopiable page 170; current class timetable; calendar for the current school year showing term dates.

Starter

- Organise the learners into pairs and give each pair a set of cards made from photocopiable page 170. Ask pairs to arrange the units of time from shortest to longest.
- Discuss the correct order, asking the learners to describe each unit of time in terms of at least one other unit (for example one hour equals 60 minutes and it also equals $\frac{1}{24}$ of a day).
- Ask the learners to write on each card an event whose duration they might measure using that unit of time. Discuss ideas, focusing on events with widely variable durations, which may be measured using a variety of units of time.

Main activities

- Hand out copies of the current class timetable and of the calendar for the current school year. Ask: *How much time will you spend studying Maths: a) today? b) this week? c) this term? d) this year?*
- The learners should be able to use the timetable to answer the first two questions, and then use their answer to question b) together with the school calendar to work out the answers to the final two questions.
- Invite answers. Discuss appropriate units of time for each answer, for example minutes may be suitable for the answer to question a) and hours for b).

- Most of the learners will probably express the answers to c) and d) in hours as well, but some learners may express them in days or weeks. If different learners have used different units for the same answer, challenge them to convert between units in order to compare their answers.
- Ask the learners to write their own questions that can be answered using the class timetable and / or the school calendar, and give them to a friend to solve.

Plenary

- Ask the learners to calculate the total amount of time they will spend at school a) today and b) this week.
- Ask the learners to give their answers and describe their methods.

Success criteria

Ask the learners:

- Can you name an event you might measure in days?
- What unit or units of time would you use to measure how long it takes to bake a cake?
- What is the name given to a period of 10 years?
- Choose a school subject. How long will you spend studying it this week?

Ideas for differentiation

Support: In the Main activity, pair these learners with a more confident partner, and / or ask them to work out the answers to the first two questions only.

Extension: Challenge these learners to calculate the amount of time they have spent studying Maths since they started school.

Units of time cards

seconds	minutes
hours	days
weeks	months
years	decades
centuries	millennia

 Cambridge Primary: Ready to Go Lessons for Maths Stage 4 © Hodder & Stoughton Ltd 2013

Length 3

Learning objectives

- Choose and use standard metric units and their abbreviations when estimating, measuring and recording length, weight and capacity. (4Ml1)
- Know and use the relationships between familiar units of length, mass and capacity; know the meaning of kilo-, cent- and milli-. (4Ml2)
- Where appropriate, use decimal notation to record measurements, e.g. 1.3 m, 0.6 kg, 1.2 l. (4Ml3)
- Interpret intervals and divisions on partially numbered scales; record readings accurately. (4Ml4)

Resources

Cards made from photocopiable pages 172 and 173; timer; rulers marked in mm and cm; metre sticks and / or tape measures marked in m and cm.

Starter

- Shuffle the cards made from photocopiable pages 172 and 173. Keep the card marked 'START' and give the rest of the cards to the learners. Some learners may need more than one card. Call out the question on the 'START' card, asking the learner with the answer to read it out and then read out the question on the bottom of their card. Continue with the answers and questions in the same way until you reach the 'END' card.
- Redistribute the cards and repeat the activity, challenging the learners to complete it in a given time limit, for example three minutes.

Main activities

- Organise the learners into pairs and distribute measuring instruments (rulers and metre sticks and / or tape measures).

- Ask the learners to choose about ten dimensions of objects in the classroom that are measurable with the ruler (for example less than 30 cm). Ask them first to estimate these lengths, and then to measure them to the nearest millimetre, recording their estimates and measurements in a table of their own devising. Both partners should make each measurement, and compare results. If there is a discrepancy, they should measure again.
- Ask the learners to repeat the process for another six dimensions greater than one metre, for example the length of a window, the height of a door or the distance between two pieces of furniture. Ask them to measure these lengths to the nearest centimetre using the metre stick or tape measure.

Plenary

- Ask the learners to convert measurements they have made in centimetres and millimetres to centimetres, for example convert 17 cm 4 mm to 17.4 cm.
- Ask the learners to convert measurements they've made in metres and centimetres to metres, for example convert 1 m 25 cm to 1.25 m.

Success criteria

Ask the learners:

- How many units of length can you name? Can you describe the relationships between them?
- Can you estimate the length / width / height of this object?
- Can you measure its length / width / height? How close was your estimate?
- Can you write the measurement using decimal notation?

Ideas for differentiation

Support: Draw a table for recording estimates and measurements for these learners to copy.

Extension: Challenge these learners to measure lengths above 1 m to the nearest millimetre and record their measurements using decimal notation, for example 1 m 25 cm 7 mm = 1.257 m.

Length follow-me cards 1

START	1000 m	50 cm	20 mm
What is 1 km in metres?	What is half a metre in centimetres?	What is 2 cm in millimetres?	What is 4000 m in kilometres?
4 km	2.5 m	2 cm	250 m
What is 250 cm in metres?	What is 20 mm in centimetres?	What is 0.25 km in metres?	What is three-quarters of a metre in centimetres?
75 cm	1 mm	500 m	5000 m
What is one-tenth of a centimetre in millimetres?	What is half a kilometre in metres?	What is 5 km in metres?	What is 0.5 cm in millimetres?
5 mm	6.3 cm	250 cm	1 m
What is 63 mm in centimetres?	What is 2.5 m in centimetres?	What is 100 cm in metres?	What is 0.2 km in metres?

 Cambridge Primary: Ready to Go Lessons for Maths Stage 4 © Hodder & Stoughton Ltd 2013

Length follow-me cards 2

200 m	3000 m	0.2 cm	10 m
What is 3 km in metres?	What is 2 mm in centimetres?	What is 1000 cm in metres?	What is 5 m in centimetres?
500 cm	1 cm	200 cm	2000 m
What is 10 mm in centimetres?	What is 2 m in centimetres?	What is 2 km in metres?	What is 1 cm in millimetres?
10 mm	1000 cm	0.01 m	10 cm
What is 10 m in centimetres?	What is 1 cm in metres?	What is 100 mm in centimetres?	What is 0.1 km in metres?
100 m	100 mm	2000 m	1000 mm
What is 10 cm in millimetres?	What is 2 km in metres?	What is 1 m in millimetres?	END

Area and perimeter 3

Learning objectives

● Draw rectangles and measure and calculate their perimeters. (4Ma1)

● Understand that area is measured in square units, e.g. cm squared. (4Ma2)

● Find the area of rectilinear shapes drawn on a square grid by counting squares. (4Ma3)

Resources

Photocopiable page 175; plain paper; rulers; set squares; squared paper.

Starter

• Hand out photocopiable page 175, which shows various rectangles drawn on a square grid.

• Revise the term 'area' (the amount of space a 2D shape covers). Ask the learners to write inside each rectangle its area in square centimetres (for example A = 30 square centimetres). Ask them to describe their method (for example counting the squares inside each rectangle).

• Revise the term 'perimeter' (the distance around the edge of a shape). Ask the learners to write inside each rectangle its perimeter in centimetres (for example P = 22 cm). Ask them to describe their method (for example counting the number of side lengths around the outside of each rectangle).

Main activities

• Using a rectangle from photocopiable page 175, introduce the 'short-cut' method of calculating the area of rectangles by multiplying the length by the width. Ask the learners to check that the answer matches the answer they got by counting the number of squares inside the rectangle.

• Introduce the 'short-cut' method of calculating the perimeter of rectangles by adding double the length to double the width. Ask the learners to check that the answer matches the answer they got by counting.

• Demonstrate how to draw a rectangle of given (whole number) dimensions on plain paper, using a ruler to measure the lengths of the sides, and a set square to ensure the corners are right angles.

• Ask the learners to draw their own rectangles onto plain paper. Ask them to challenge a friend to calculate the perimeter and area of each rectangle.

Plenary

• On the board, draw diagrams of rectangles (not on a squared grid). Label each rectangle with a whole number length and width, for example 4 m by 3 m, 9 cm by 2 cm, 12 m by 1 m, 6 cm by 7 cm.

• Challenge the learners to calculate the perimeter and area of each rectangle.

Success criteria

Ask the learners:

● On squared paper, draw a rectangle with an area of 28 square centimetres. What is its perimeter?

● On squared paper, draw a rectangle with a perimeter of 30 cm. What is its area?

● Imagine a rectangle that is 4 cm long and 9 cm wide. What is its area? How did you work it out?

● Imagine a rectangle that is 6 cm long and 3 cm wide. What is its perimeter? How did you work it out?

Ideas for differentiation

Support: For the final Main activity, ask these learners to draw their rectangles on squared paper.

Extension: Challenge these learners to devise a method for finding the area of right-angled triangles if you know the length of their perpendicular sides (those at right angles to one another).

Rectangles

A
B
C
D
E
F
G

Mass 3

Learning objectives

- Choose and use standard metric units and their abbreviations when estimating, measuring and recording length, weight and capacity. (4Ml1)
- Know and use the relationships between familiar units of length, mass and capacity, know the meaning of kilo-, cent- and milli-. (4Ml2)
- Where appropriate, use decimal notation to record measurements, e.g. 1.3 m, 0.6 kg, 1.2 l. (4Ml3)
- Interpret intervals and divisions on partially numbered scales; record readings accurately. (4Ml4)

Resources

A set of seven opaque containers (e.g. cereal boxes filled with sand and taped closed) with the following masses, each container labelled clearly with its mass: $\frac{3}{10}$ kg, $\frac{1}{2}$ kg, $\frac{1}{4}$ kg, $\frac{3}{4}$ kg, $\frac{1}{10}$ kg, $\frac{7}{10}$ kg $\frac{1}{5}$ kg; mass measuring instruments suitable for measuring masses up to 1 kg (e.g. pan balances, spring scales, kitchen scales); photocopiable page 177; bathroom scales; calculators.

Starter

- Display the set of seven containers, asking the learners to order them from lightest to heaviest.
- Ask the learners to write the mass of each container in grams. (The gram equivalents should enable the learners to confirm whether they have ordered the containers correctly.)
- Confirm the gram equivalents by dividing the class into seven groups and give each group a container to weigh and a suitable measuring instrument to weigh it with.
- Finally, confirm the correct order of the containers ($\frac{1}{10}$ kg, $\frac{1}{5}$ kg, $\frac{1}{4}$ kg, $\frac{3}{10}$ kg, $\frac{1}{2}$ kg, $\frac{7}{10}$ kg, $\frac{3}{4}$ kg).

Main activities

- Distribute photocopiable page 177 and discuss the pictures. Explain that a maximum load is the greatest mass that something can carry safely at one time.

- Organise the learners into groups. Give each group a set of bathroom scales and at least one calculator. Ask the learners in each group to choose one of the pictures and calculate how many children of their age could use it safely at any one time.
- Ask the learners what information they need to collect (for example roughly how much one child of their age weighs). Discuss possible ways of doing this (for example choose a 'typical' child from the class and find their mass).
- Ask the learners what calculation they will need to do in order to answer the question (divide the maximum load of the machine by the mass of a child). Ask the learners to use the calculators for this, and to check their answers by carrying out the inverse operation (for example multiplication) using the calculator.

Plenary

- Ask the learners to share their answers.
- Ask: *Should your answer be a whole number?* (Yes, because you're counting children.) *Should you round up or down? Why?* (Down, otherwise the maximum load will be exceeded.)

Success criteria

Ask the learners:

- Can you order these masses from lightest to heaviest? $\frac{3}{4}$ kg, $\frac{1}{4}$ kg, $\frac{1}{2}$ kg, $\frac{3}{10}$ kg
- Can you give each mass from the previous question in grams?
- Can you measure my mass in kilograms?
- About how many people with a similar mass to mine could safely use the rope bridge at the same time?

Ideas for differentiation

Support: Group these learners together, and work with them, helping them to measure each other's masses and calculate their average mass.

Extension: Challenge these learners to find answers for all the machines pictured on photocopiable page 177.

Maximum loads

Choose one of these. Work out how many children in your class could use this safely at one time.

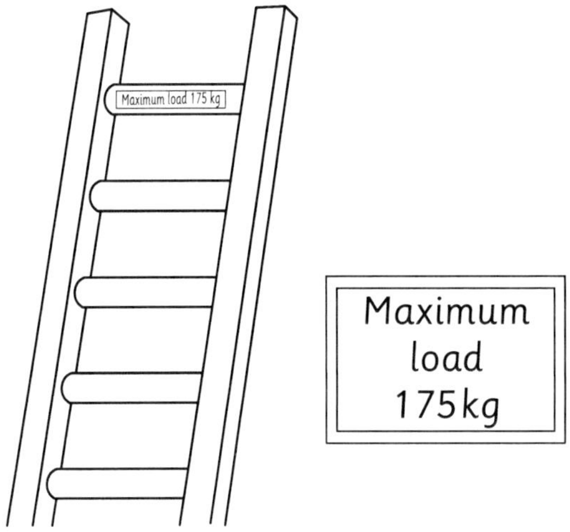

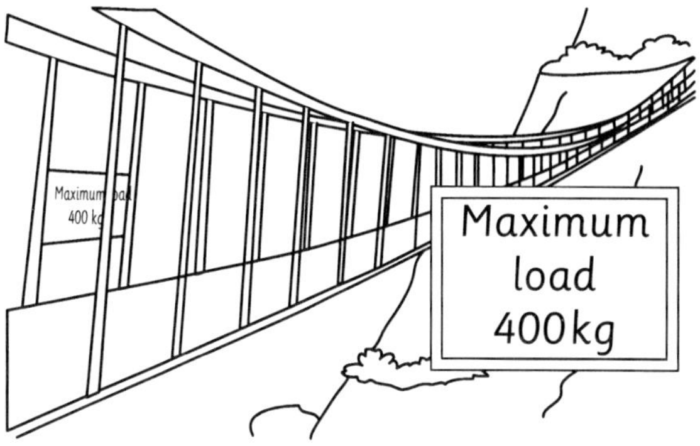

Maximum
load
175kg

Maximum
load
400kg

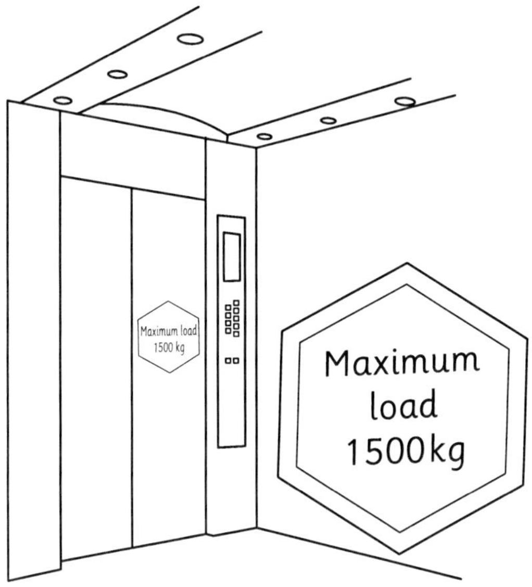

Maximum
load
1500kg

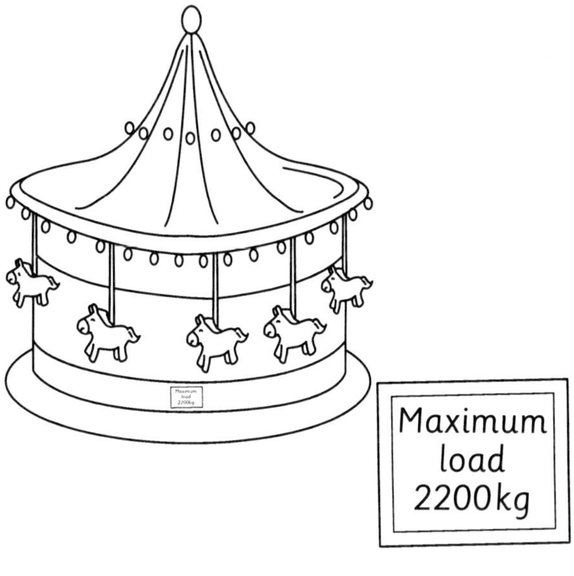

Maximum
load
2200kg

Capacity 3

Learning objectives

- Know and use the relationships between familiar units of length, mass and capacity, know the meaning of kilo-, cent- and milli-. (4Ml2)
- Where appropriate, use decimal notation to record measurements, e.g. 1.3 m, 0.6 kg, 1.2 l. (4Ml3)
- Understand everyday systems of measurement in length, weight and capacity and time, and use these to solve simple problems as appropriate. (4Pt2)
- Make up a number story for a calculation, including in the context of measures. (4Ps1)

Resources

Photocopiable page 179.

Starter

- On the board, write the following capacities: $\frac{1}{2}$ litre, $\frac{6}{10}$ litre, $\frac{1}{4}$ litre, $\frac{4}{5}$ litre, $\frac{3}{4}$ litre, $\frac{1}{10}$ litre.
- Ask the learners to order the capacities from least to greatest.
- Ask them to write each capacity in millilitres. (The millilitre equivalents should let the learners see whether they have ordered the capacities correctly.)
- Confirm the correct order and the millilitre equivalents: $\frac{1}{10}$ litre (100 ml), $\frac{1}{4}$ litre (250 ml), $\frac{1}{2}$ litre (500 ml), $\frac{6}{10}$ litre (600 ml), $\frac{3}{4}$ litre (750 ml), $\frac{4}{5}$ litre (800 ml).
- Finally, ask the learners to express each capacity in litres using decimal notation, for example $\frac{1}{10}$ litre = 0.1 litres.

Main activities

- Distribute photocopiable page 179. Read through the first problem together with the learners and ask them what calculation they need to do ($\frac{1}{4}$ litre + $\frac{1}{5}$ litre + $\frac{1}{2}$ litre). Ask: *Is it easy to do this calculation?* (No.) *Why not?* (Because the fractions have different denominators.) Ask: *How could you make the calculation easier?* (Write each capacity without a fraction.)

- Ask the learners to write each capacity without a fraction. Discuss the need for the three capacities to be expressed using the same unit (for example all in millilitres, or all in litres using decimal notation). The learners may make different choices here: some may prefer to deal with the whole numbers produced by converting to ml, and some may prefer to deal with the smaller (decimal) numbers produced by converting to litres.
- Once the learners have successfully converted the capacities, ask them to carry out the calculation.
- Ask them to solve the rest of the problems on photocopiable page 179.

Plenary

- Ask selected learners to give the answers to the word problems on photocopiable page 179. Ask the other learners to express each answer using different units.
- Ask volunteers to read out the number stories they wrote.

Success criteria

Ask the learners:

- Choose one of the problems on photocopiable page 179 that you have already answered. How did you work out the answer?
- What answer did you get?
- Can you write your answer in a different way?
- Can you make up a number story to go with this calculation: 5 × 750 ml = 3.75 litres?

Ideas for differentiation

Support: Group these learners together and guide them through an extra problem. Ask them to miss out questions 5 and 6.

Extension: Ask these learners to make up word problems based on their own ideas instead of the questions given on photocopiable page 179.

Name: _____

Capacity problems 2

1. I have three containers. One holds $\frac{1}{4}$ litre, one holds $\frac{1}{5}$ litre

    and one holds $\frac{1}{2}$ litre. What is the total capacity of the

    three containers?

    _250 + 200 + 500 = 950 ml_
    1600   1750

2. A red jug holds $1\frac{6}{10}$ litres. A blue jug holds $1\frac{3}{4}$ litres.

    a) Which jug holds more? _____

    b) How much more does it hold? _____

3. A recipe for mushroom soup uses $1\frac{1}{5}$ litres of milk,

    $\frac{3}{4}$ litre of chicken stock and $\frac{3}{10}$ litre of cream.

    How much liquid does the recipe use altogether?

    _1200 + 750 + 300 = 2250_

4. I have five bottles of apple juice. Each bottle contains $1\frac{1}{4}$ litres.

    How much apple juice do I have altogether?

    _6.25 l_

5. To fill a paddling pool with water, I pour in

    ten $4\frac{1}{2}$ litre buckets and ten $5\frac{1}{4}$ litre buckets.

    How much water is in the paddling pool?

    _45 + 52.5 = 97.5_

6. Write a number story on the back of this page to go with each of
    these calculations:

    a) $5\frac{7}{10}$ litres − $2\frac{1}{2}$ litres

    b) $2\frac{1}{4}$ litres + $1\frac{1}{2}$ litres

    c) $3 \times 5\frac{1}{2}$ litres

    Work out the answers to your number stories.

Unit assessment

- How would you write seventeen minutes to three o'clock in the afternoon on a digital clock?
- What will the date be in three weeks' time?
- What unit or units of time would you use to measure how long it takes to read a book?
- What are the units of mass? Can you describe the relationship between them?

- Can you measure the width of the board to the nearest centimetre? Can you write the measurement using decimal notation?
- Can you order these capacities from least to greatest? $1\frac{1}{2}$ litres, $1\frac{2}{5}$ litres, $1\frac{1}{4}$ litres, $1\frac{3}{4}$ litres, $1\frac{6}{10}$ litres

Summative assessment activities

Observe the learners while they take part in these activities. You will quickly be able to identify those who appear to be confident and those who may need additional support.

Animal problems

This activity assesses the learners' ability to understand everyday systems of measurement in mass, and to use these to solve simple problems.

You will need:

A list of the average masses, in kg, of about a dozen different species of animal, some lighter than humans and some heavier, for example: duck – 1 kg; cat – 4 kg; red fox – 9 kg; wolf – 30 kg; human – 70 kg; goat – 77 kg; gorilla – 150 kg; lion – 200 kg; polar bear – 600 kg; giraffe – 1000 kg; hippopotamus – 1800 kg; African elephant – 4900 kg.

What to do

- Give the learners a copy of the list of average masses of various animals. Ask them to write word problems based around these measurements. Challenge them to write at least one problem for each operation (addition, subtraction, multiplication and division).
- Ask the learners to give their problems to a friend to solve.

Rectangle areas

This game assesses the learners' ability to draw rectangles and calculate their perimeters, and to find the area of rectilinear shapes drawn on a square grid by counting squares.

You will need:

Sets of 12 'perimeter cards' labelled 14 cm, 16 cm, 18 cm, 20 cm ... 36 cm; squared paper.

What to do

- Organise the learners into groups. Give each group a set of perimeter cards and plenty of squared paper. Ask the learners to shuffle the perimeter cards, place the cards face down in a pile and turn over the top card. The players should each draw a rectangle on squared paper, which must have the given perimeter. Players who draw the rectangle with the largest area score a point.
- Repeat the game, with the aim of making a rectangle with the smallest area.

Distribute photocopiable page 181. Ask the learners to read the questions and write the answers. They should work independently.

Name: _____

Looking at measures

1. Draw the hands on each clock face to show the time written underneath it.

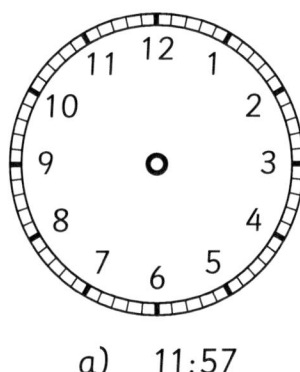
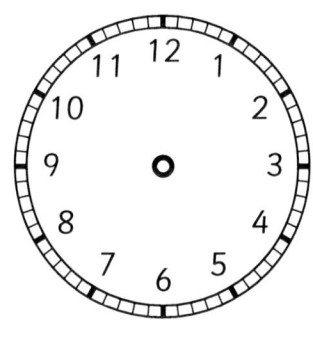
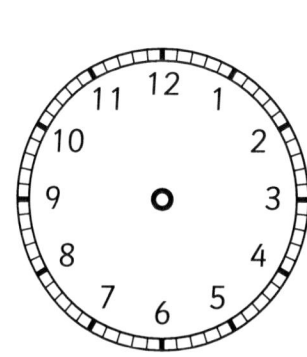

   a)   11:57   b)   sixteen minutes to three   c)   4:22

2. How much milk is in this jug? _____

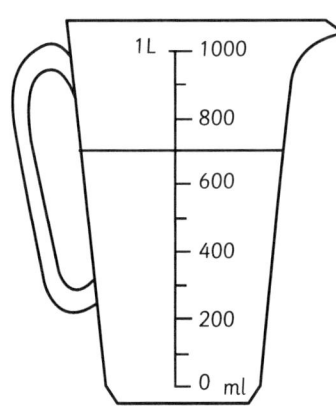

3. Imagine a rectangle that is 5 cm long and 8 cm wide.

   a)   What is its area? _____

   b)   What is its perimeter? _____

4. What is 20 metres written in centimetres? _____

5. Some friends start playing a game at 4:20 p.m.

   The game takes 1 hour 43 minutes.

   What time do the friends finish playing?

   _____

6. I have three containers. One holds $\frac{3}{4}$ litre, one holds $\frac{1}{2}$ litre and one

   holds $\frac{2}{5}$ litre. What is the total capacity of the three containers?

   750 + 500 + 600 = 1650

Handling data 5

Learning objectives

● Answer a question by identifying what data to collect, organising, presenting and interpreting data in tables, diagrams, tally charts, frequency tables, pictograms and bar charts. (4Dh1)

● Use ordered lists and tables to help to solve problems systematically. (4Ps5)

● Explain methods and reasoning orally and in writing; make hypotheses and test them out. (4Ps9)

Resources

Photocopiable page 183.

Starter

• Display a copy of photocopiable page 183. Ask the learners to name each type of diagram (bar chart, frequency table, pictogram and tally chart).

• Ask the learners questions requiring them to interpret the data and compare the different representations, for example:
  • *How many children like Fridays? Which diagram or diagrams show this most clearly? Why?*
  • *Which day of the week is half as popular as the most popular day? Which diagram or diagrams show this most clearly? Why?*
  • *How many children are in the class altogether? Which diagram did you use to work out your answer? Why?*

• Keep photocopiable page 183 on display throughout the rest of the lesson.

Main activities

• On the board, write four or five questions about the class that can be answered in a short time either by taking a survey or by making measurements, for example:
  • How big are our families?
  • Who has the widest arm span?
  • How many of us were born in each month?
  • What time do we get up on a school day?
  • What is our favourite fruit?

• Ask the learners what data they would need to collect in order to answer each question, how they might collect it, and what type of diagram they would draw and why.

• Ask the learners to choose one of the questions on the board. Group the learners according to their choice of question. (Several small groups working on a question works better than one large group.) Ask the learners to collect the data they need, and present it in the form of a diagram of their choice.

Plenary

• Ask a volunteer from each group to describe which question they chose, what data they collected and how they collected it. Ask them to show the diagram their group drew.

• Ask each group: *Do you have enough information to answer your question? If you do, what is the answer? If you don't, what extra data could you collect that might help you answer the question?*

Success criteria

Ask the learners:

● What question are you trying to answer?

● What do you think the answer to the question will be? Why?

● What data are you collecting? How will this data help you to answer the question you're trying to answer?

● What does the data in this diagram tell you?

Ideas for differentiation

Support: To support these learners, organise mixed-ability groupings in the final Main activity.

Extension: Challenge these learners to devise another question of their own to investigate (perhaps this could lead on from the original question they chose).

Same data, different diagrams

A group of school children asked their classmates the question:

What is your favourite day of the school week?

Each member of the group drew a different type of diagram to represent the data they collected.

Ami's diagram

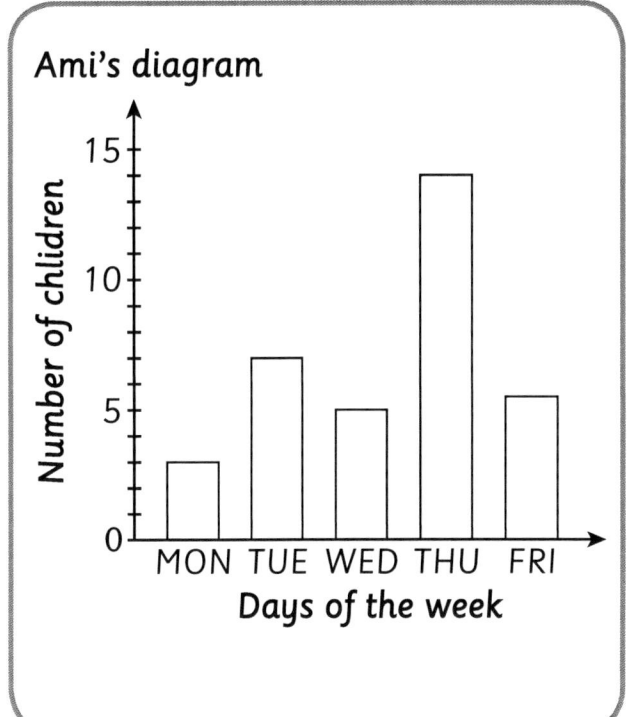

Lilly's diagram

DAY	NUMBER OF CHILDREN
Monday	3
Tuesday	7
Wednesday	5
Thursday	14
Friday	6

Sameera's diagram

DAY	NUMBER OF CHILDREN
MON	☺ ☺
TUE	☺ ☺ ☺ ☺
WED	☺ ☺ ☺
THU	☺ ☺ ☺ ☺ ☺ ☺ ☺
FRI	☺ ☺ ☺

Key: ☺ = 2 children

Sofia's diagram

DAY	NUMBER OF CHILDREN
Monday	III
Tuesday	ЖЖ II
Wednesday	ЖЖ
Thursday	ЖЖ ЖЖ IIII
Friday	ЖЖ I

Handling data 6

Learning objectives

● Compare the impact of representations where scales have different intervals. (4Dh2)

● Explain methods and reasoning orally and in writing; make hypotheses and test them out. (4Ps4)

Resources

Counting stick; photocopiable page 185; coloured pencils; squared paper.

Starter

• Using the counting stick held vertically, practise counting on from 0 (at the bottom of the stick) in 20s and back again.

• Point to each division on the counting stick in a random order, asking the learners to say each number aloud, for example if you point to the third division from the bottom, the learners should say 'sixty'.

• Repeat the activity for counting in 10s, 5s and 2s.

Main activities

• Distribute photocopiable page 185, coloured pencils and squared paper. Ask the learners questions about the data, for example: *Can you see any patterns in the data?* (The average test scores are going up.) *By how many points does the average test score rise over the course of the school year?* (15.) *Between which two months does the average test score increase the most?* (January and March.)

• Ask the learners to draw two bar charts showing the data in the frequency table, each with a different scale on the vertical axis. To do this, they should use the pre-drawn axes on photocopiable page 185.

• After the learners have drawn the two bar charts, ask them to imagine that they are the class teacher and they want to emphasise the improvement the class have made over the course of the year. Ask: *Which bar chart would you choose? Why?*

• Ask the learners to devise their own scale that would be even better at emphasising the improvement the class have made, and draw a bar chart using that scale on squared paper.

Plenary

• Ask the learners to share and compare the bar charts they have drawn using their own scale. Discuss which scale is the most effective at emphasising the class's improvement in test scores, and why.

• Ask the learners to say which bar they would choose if they wanted to suggest that the children's test scores didn't improve very much, and explain why.

Success criteria

Ask the learners:

● How does using a different scale on the vertical axis change the way a bar chart looks?

● What scale have you chosen to draw the third bar chart? How did you decide which scale to use?

● If you want to make differences in data values look smaller, where should your scale start? Should each division represent a small number or a large number?

● If you want to make differences in data values look bigger, where should your scale start? Should each division represent a small number or a large number?

Ideas for differentiation

Support: Group these learners together and work with them in the final Main activity, helping them to draw a bar chart with a suitable scale.

Extension: Ask these learners to present the data they collected in the previous lesson in the form of a bar chart, choosing a scale to emphasise a particular aspect of the data.

Same data, different scales

The frequency table below shows the average test score in Maths in a certain class over the course of a school year.

Month	September	November	January	March	May	July
Average test score in Maths (out of 100)	58	61	65	70	72	73

Draw two bar charts below to show the data in the frequency table. You will notice that each has a different scale on the vertical axis.

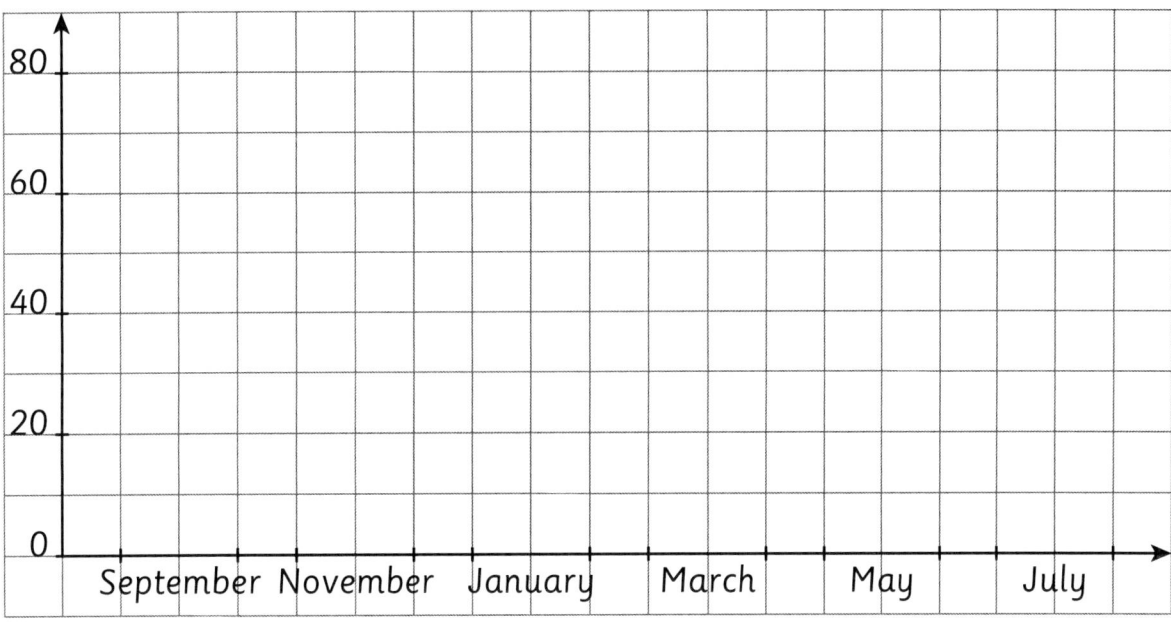

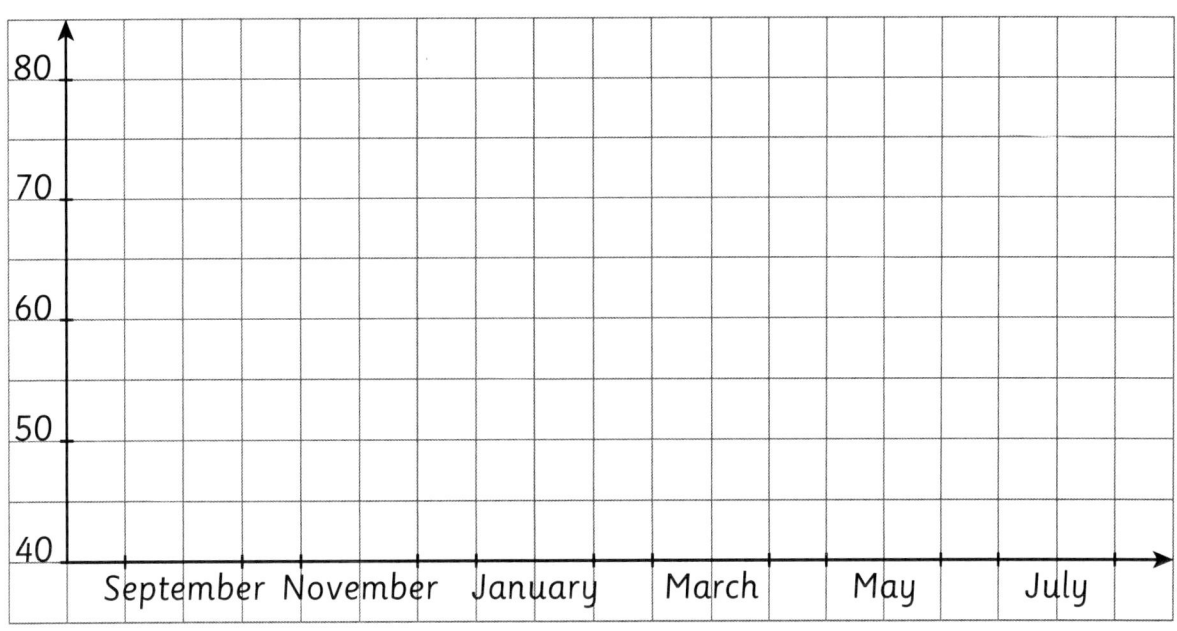

Handling data 7

Learning objectives

- Use Venn or Carroll diagrams to sort data and objects using two or three criteria. (4Dh3)
- Explain methods and reasoning orally and in writing; make hypotheses and test them out. (4Ps9)

Resources

Photocopiable page 187; A3 paper.

Starter

- Draw a four-region Carroll diagram on the board (see photocopiable page 187 for an example). Label the columns 'less than 10' and 'not less than 10' and label the rows 'multiple of 4' and 'not a multiple of 4'.
- Ask the learners to copy the diagram and write each whole number from 0 to 20 in the correct region of the diagram. Share answers.
- Repeat the activity, relabelling the columns 'odd' and 'even' and relabelling the rows 'one syllable' and 'more than one syllable'. Share answers.

Main activities

- Display the Carroll diagram on photocopiable page 187. Ask questions about the data, for example: *What set of data has been sorted in this diagram?* (Days of the week.) *Which day has more than seven letters and doesn't contain the letter t? Do most days contain the letter t?*
- Revise Venn diagrams using the Venn diagram on photocopiable page 187. Ask the learners questions, for example: *What set of data has been sorted in this diagram?* (Food preferences of a group of people.) *Which food is the most popular? Who likes chocolate but doesn't like ice cream or pizza? What can you tell me about Oscar?*
- Organise the learners into groups and hand out A3 paper. Ask the learners to choose a set of data (for example numbers, words, shapes or facts about the members of the group) and sort it according to their own criteria, using a Carroll or a Venn diagram. Ask them to sort a second set of data using the other type of diagram.

Plenary

- Ask groups to show and explain the Carroll and Venn diagrams they have drawn.
- Draw a three-circle Venn diagram or a four-region Carroll diagram on the board. Write in data (for example shapes or numbers), but do not write in labels for the circles / boxes. Ask the learners to suggest what the labels might be.

Success criteria

Ask the learners:

- Where does this shape / number / word / person belong in this Carroll diagram? Why?
- Where does this shape / number / word / person belong in this Venn diagram? Why?
- Can you draw a three-circle Venn diagram to sort these shapes?
- Can you draw a four-box Carroll diagram to sort these numbers?

Ideas for differentiation

Support: Group these learners together and help them to devise the sorting criteria they are going to use, ensuring that each pair of criteria in the Carroll diagram covers all possible answers.

Extension: Ask these learners to suggest other data / objects to sort using a Venn or Carroll diagram.

Carroll and Venn diagrams

A Carroll diagram

	contains the letter t	does not contain the letter t
more than 7 letters	Thursday Saturday	Wednesday
not more than 7 letters	Tuesday	Monday Friday Sunday

A Venn diagram

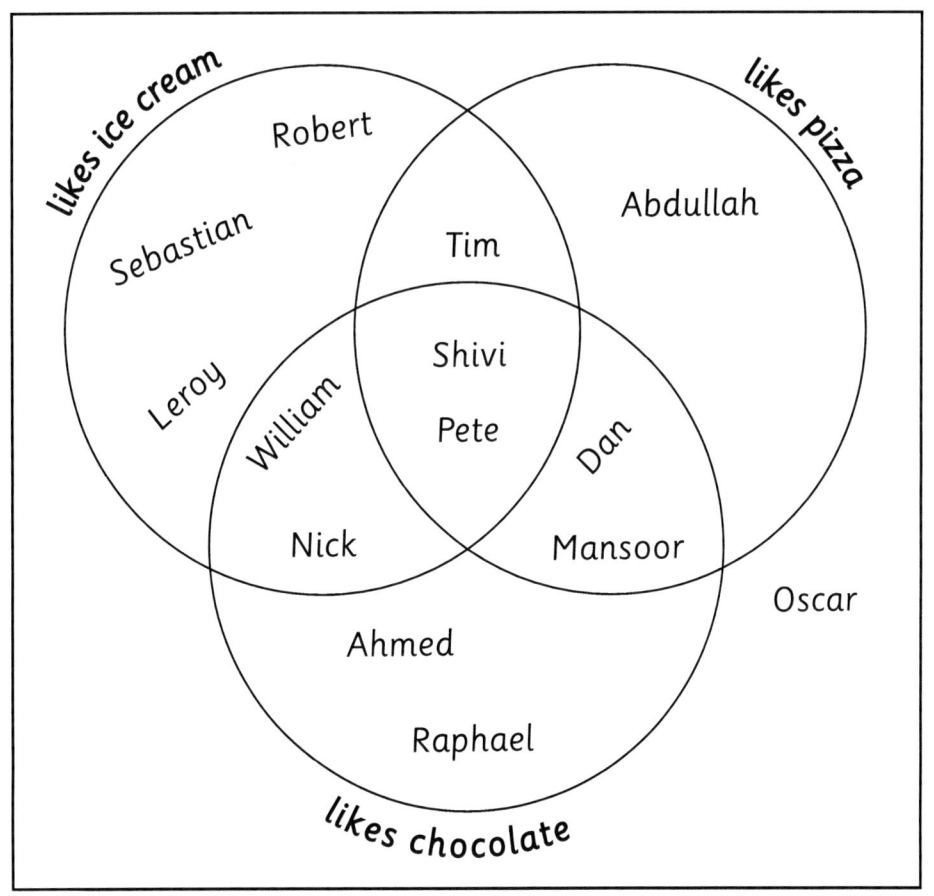

Handling data 8

Learning objectives

● Answer a question by identifying what data to collect, organising, presenting and interpreting data in tables, diagrams, tally charts, frequency tables, pictograms and bar charts. (4Dh1)

● Use ordered lists and tables to help to solve problems systematically. (4Ps5)

● Explain methods and reasoning orally and in writing; make hypotheses and test them out. (4Ps9)

Resources

Cards made from photocopiable page 189; A3 paper; marker pens; photocopiable page 190; squared paper; rulers; coloured pencils.

Starter

- On the board, write a list of seven whole-number data values, for example 5, 8, 15, 2, 7, 12, 4.

- Organise the learners into pairs and give each pair a piece of A3 paper, a marker pen, and a card made from photocopiable page 189, featuring the name of a type of data diagram.

- Ask the learners to sketch a large diagram on the paper using the marker pen. The diagram should match the type on the card, and should represent the data values on the board.

- Ask the learners to imagine what the data might be about, and indicate this imaginary context in the labelling on their diagram.

- Ask the learners to show their diagrams and describe the context they invented. Compare and contrast different diagrams of the same type.

Main activities

- Display a copy of photocopiable page 190, which poses a variety of questions that can be answered by collecting data. For each question, ask the learners to predict the answer and explain their reasoning. Ask: *What data would you need to collect in order to answer the question? How might you collect it?*

- Ask: *How might you present the data after you have collected it?* Revise tally charts, frequency tables, bar charts and pictograms as ways of organising and presenting data.

- Organise the learners into groups. Ask each group to choose one of the questions on photocopiable page 190 to answer. Ask each group to decide for themselves how they will collect the data and how they will present it.

Plenary

- Ask a representative from each group to say which question they chose, and describe how they collected the data.

- Ask them to display the chart or graph they drew, and explain what it shows.

Success criteria

Ask the learners:

● What question did you choose?
● What data did you collect and how did you collect it?
● How did you organise and present the data?
● What does the data you collected show?

Ideas for differentiation

Support: Organise these learners into mixed-ability groups, so that the less-able learners are supported by other group members.

Extension: Ask these learners to ask and answer their own follow-up question to the initial question they investigated.

Data diagram cards

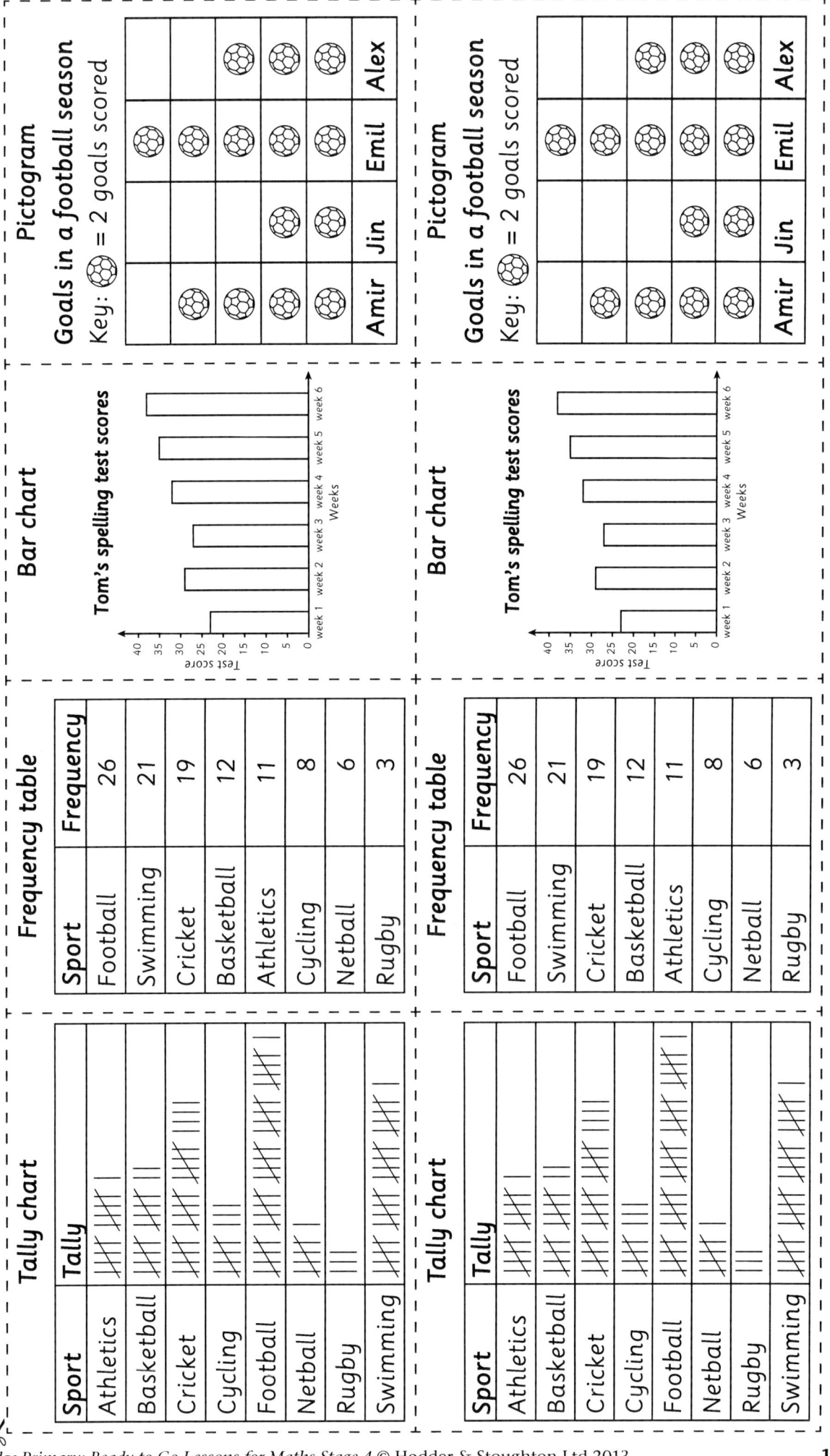

Tally chart

Sport	Tally
Athletics	卌 卌 I
Basketball	卌 卌 II
Cricket	卌 卌 卌 IIII
Cycling	卌 III
Football	卌 卌 卌 卌 卌 I
Netball	卌 I
Rugby	III
Swimming	卌 卌 卌 卌 I

Frequency table

Sport	Frequency
Football	26
Swimming	21
Cricket	19
Basketball	12
Athletics	11
Cycling	8
Netball	6
Rugby	3

Bar chart

Tom's spelling test scores

Pictogram

Goals in a football season

Key: ⚽ = 2 goals scored

| | | Amir | Jin | Emil | Alex |

Data handling questions

Choose one of these questions to answer.

1. Can the people who can jump the furthest also jump the highest?

2. Are some people better at identifying smells than others?

3. What is the average amount of pocket money in our class?

4. What is the most popular school subject?

5. What is the most common age of class members in years and months?

6. Who is the most popular author in our class?

 Cambridge Primary: Ready to Go Lessons for Maths Stage 4 © Hodder & Stoughton Ltd 2013

Unit assessment

Questions to ask

- What did you find out from the question you investigated in the last lesson?
- What type of diagram did you choose to represent your data? Why did you choose this type of diagram?
- Can you draw a Venn diagram or a Carroll diagram to sort these 2D shapes?
- How does the scale you choose for a bar chart affect the way the data appears?

Summative assessment activities

Observe the learners while they take part in these activities. You will quickly be able to identify those who appear to be confident and those who may need additional support.

Drawing a bar chart

This activity assesses the learners' ability to interpret frequency tables and draw bar charts.

You will need:

Squared paper; rulers; pencils; erasers; coloured pencils.

What to do

- Draw a frequency table on the board. You could use or adapt the table below.

Toys in Class 4	
Toy	**Number of children**
Bicycle	12
Swing	9
Scooter	5
Rollerskates	6
Trampoline	3
Skateboard	7
Kite	8

- Ask the learners to draw a bar chart of the data shown in the frequency table.

Interpreting data

This activity assesses the learners' ability to interpret data in bar charts.

You will need:

Bar charts drawn by the learners in the previous assessment activity 'Drawing a bar chart' (see left). If any learners failed to draw a successful bar chart in the first assessment activity, give them a copy of another learner's bar chart for this activity.

What to do

- Ask the learners questions requiring them to interpret the data in the bar chart they drew, for example:
  - *Which is the commonest toy in Class 4?*
  - *How many more children have swings than have trampolines?*
  - *How many children have either a scooter or rollerskates?*
  - *Can you tell how many children there are in Class 4?* (No.) *Why not?* (Because some children may have more than one type of toy, and so be counted more than once. Also, there may be some children who do not have any of these types of toys.)
  - *Can you tell how many kites are owned by the children in Class 4?* (No.) *Why not?* (Because the data only tells you how many children have kites, not how many kites they have. Some children could have more than one kite.)
- Depending on the size of the group you are assessing and the ability of the learners in it, you may want the learners to give their responses orally, or record them in writing.

Written assessment

Distribute photocopiable page 192. Ask the learners to read the questions and write the answers. They should work independently.

Name: _____

Looking at data

The frequency table below shows the number of sunglasses sold in one week at a sunglasses kiosk.

Day	Mon	Tues	Wed	Thu	Fri	Sat	Sun
Number of sunglasses sold	35	25	10	20	5	45	40

1. Draw a pictogram showing the data in the table. Remember to include a key.

2. Write four questions about the data in the table.
    Give them to a friend to answer.

    _____

    _____

    _____

    _____

Cambridge Primary: Ready to Go Lessons for Maths Stage 4 © Hodder & Stoughton Ltd 2013